Kick the Tires and Light the Fires

Kick the Tires and Light the Fires

My Life as a Naval Aviator, FAA Test
Pilot, and Aviation Consultant

by DAVID PAUL WEST
as told to Ron Martz

Deeds Publishing | Athens

Published by Deeds Publishing in Athens, GA
www.deedspublishing.com

Printed in The United States of America.

Cover photo by Jaron Berman.
Cover and interior design by Deeds Publishing.

ISBN 978-1-961505-17-9

Books are available in quantity for promotional or premium use. For information, email info@deedspublishing.com.

First Edition, 2024

10 9 8 7 6 5 4 3 2 1

To our children:
David, Travis, Shannon, Erin, and Jake

and our grandchildren:
Stephanie and David 3rd, Jess and George, Hannah and Toby, Thomas and Ryan, and Jupiter

CONTENTS

GLOSSARY

Ab initio trainer—An aircraft for pilots just learning to fly

ACO—Aircraft Certification Office

ADF—Automatic Direction Finder

AGM—Air to Ground Missile

ATC—Air Traffic Control

Angels—Navy term for altitude in thousands of feet, as in "Angels Two" is 2,000 feet in altitude

Angle of Attack (AOA)—Angle between the plane of the aircraft's wing and the oncoming airflow

Ball—Also referred to as "Meatball," is the amber light in the Fresnel lens on the port side of the aircraft carrier near the stern that the pilot uses as a guide to land. If the pilot is coming in too high, the ball appears at the top of the screen. If too low, the ball is at the bottom of the screen and turns red, usually necessitating a wave-off by the Landing Signal Officer.

Bilge—to fail a test or course at the Naval Academy

Black Shoe Navy—Everybody in the Navy but aviators

Boat—What Naval aviators call the aircraft carrier; for everyone else in the Navy, it's a ship

Bolter—When an aircraft's tailhook misses one of the four wires on the deck and the pilot must hit the gas and get back into the pattern to make another attempt to land

Blue trampoline—Bunks at the Naval Academy because of the blue blankets

Brown Shoe Navy—Naval aviators

Bulkhead—Wall

CAS—In the military, Close Air Support. In civilian aviation, Collision Avoidance System.

Cover—A hat or cap

CTF—Carrier Task Force

CVW—Carrier Air Wing

Deck—Floor

DER—Designated Engineering Representative

Detailers—Also known as cadre, they are the upperclassmen whose job is to indoctrinate Plebes when they arrive at Annapolis for Plebe Summer

Dixie Cup—the round, white cap (or cover) associated with Navy enlisted; Plebes at the Naval Academy are recognizable by the blue rim on their Dixie cups

DMZ—Demilitarized Zone

DOA—Designated Option Authorization

ECM—Electronic Countermeasures

EW—Electronic Warfare

FAA—Federal Aviation Administration

FARs—Federal Aviation Regulations

Feet Dry—Flying over land

Feet Wet—Flying over water

FMLP—Field Mirror Landing Practice

FRS—Fleet Replacement Squadron

G's or G-forces—The gravitational force to which an object is subjected when it is accelerated

ICAO—International Civil Aviation Organization

IFR—Instrument Flight Rules

JIFDATS—Joint In-Flight Data Acquisition Transmission System

Jimmy Legs—Civilian gate guards at the Naval Academy

JPATS—Joint Primary Aircraft Training System

LSO—Landing Signal Officer on an aircraft carrier; also referred to as "Paddles"

MCAS—Maneuvering Characteristics Augmentation System

MIA—Missing in Action

Mother B—Bancroft Hall, home to the Brigade of Midshipmen at Annapolis

NAS—Naval Air Station

NATO—North Atlantic Treaty Organization

NFO—Naval Flight Officer

NVA—North Vietnamese Army

O Club — Officers Club

Oiler — Navy resupply ship

Overhead — Ceiling

Over the Wall — Leaving the Naval Academy for short periods of time without authorization

Paddles — Landing Signal Officer on an aircraft carrier

Plebe — First-year Midshipman

POW — Prisoner of War

RA — Resolution Advisory

RAG — Replacement Air Group

R&R — Rest and Recuperation

Recces — Reconnaissance flights

Reef Points — The 200-plus-page manual for Midshipmen at the Naval Academy

RFP — Request for Proposal

RIO — Radar Intercept Officer, the back seater in an F-4B who handled most of the electronics

SAM — Surface to Air Missile

SERE — Survival, Evasion, Resistance, and Escape

Sierra Hotel — Also "Shit Hot," sometimes abbreviated in writing as S/H, it is a pilot's way of saying "I agree wholeheartedly." Or, "That has my greatest admiration."

Spooning — Shaking someone's hand

STOL — Short Takeoff and Landing

TA — Traffic Alert

TCAS — Traffic Collision Avoidance System

TIA — Type Inspection Authorization

TIR — Type Inspection Report

UPRT — Upset Prevention and Recovery Training, often referred to only as "upset training," is designed to teach pilots how to recover from situations that could lead to loss of control of the aircraft in flight.

V Speeds — Designated speeds for specific flight conditions as established by aircraft manufacturers and designers during testing for certification. They are specific to aircraft type and model.

VFR — Visual Flight Rules

FOREWORD

Writing his memoirs became Dave's focus in early 2019. Beginning in May of that year, he met with Ron Martz, a journalist who agreed to take Dave's stories and compile them into a readable format. Anyone who knew Dave knew he was a *raconteur extraordinaire*; he could keep his audience spellbound for hours as he recounted story after story of his life in the U.S. Navy and as a test pilot.

I have told many people over the years, "God broke the mold after he made Dave West." He *was* a singular man. I had never met anyone like him when we first met on August 1, 1986, and I doubt I will ever meet his equal.

In Dave's senior year of high school, the yearbook staff chose to include a quote from Shakespeare beside each senior's photo. The quote beside Dave's picture was taken from *Hamlet*: "He was a man; take him for all in all. I shall not look upon his like again." Truer words were never spoken.

Although he was not enamored of the Naval Academy when he was there as a Midshipman, his attitude toward the Academy changed as the years passed; in the end, he was grateful for his years at the Academy because it led to his career in aviation. In fact, he could still quote many of the "Table Salt" sayings from the Academy's *Reef Points*.

When I met Dave, he was a black-and-white kind of man with very little "gray area" in his viewpoints. As you read his memoir, you will hear him say over and over, "There's a right way and a wrong way to do (fill in the blank)." He always tried to do things the right way. As he grew older, however, he did allow a little gray to creep into his attitudes about certain circumstances in life.

He could be gruff with people he encountered professionally, but he was a very tender, sensitive man deep down inside. He was a loyal family man; family meant everything to him, especially when the children became adults and we could relate to them on an equal footing. Whenever one came to visit, the deck of cards always came out, and we'd play Hearts. Dave never lost his competitive edge.

In May 2021 Dave fell in our home, breaking a hip. Following surgery to repair the break, he suffered a stroke. He never recovered and passed away on July 22, 2021. His passing left a huge hole in my heart.

In a nutshell, he was my hero and my champion. He loved deeply and loved fiercely.

* * * * *

After his death was announced, I heard from many people who had worked with Dave. Below are just a few examples of the messages I received.

Denny Moore, Dave's friend from Northwestern Prep, the Naval Academy, and a fellow naval aviator wrote:

"As you know, Dave and I were very close friends during our Navy years, and I have very vivid memories of those times. Dave went on to have a great career in aviation. I know you are justly proud of his accomplishments."

From Randy Gaston:

"Mrs. West, I heard of Dave's passing and had to send my regards. I was at the ACO from 1991-1994. Dave gave me my job there, and it turned out to be a turning point in my life. Dave hired me but had to call shortly after the offer and tell me that the FAA had wanted another candidate. I was at Edwards at the time working for Northrop as a test pilot on the B2 test program. We bought a house up in Tehachapi and settled in, and about 6 months after Dave's call to tell me he had to withdraw his offer, he called again and said the candidate did not work out and wanted to know if I was still interested. I said yes. I wanted a change from military test programs. The first kindness was Dave loaning me the brown car (loaner for when the kids visited). Our car was being shipped at the time. I was a very early riser and would be at my desk when Dave opened his office. He would get his coffee and come by my desk, and we would start to solve the world's problems. No one else was there. I enjoyed those moments. Dave could look intimidating when he wanted to, but it was only a show. His heart was always easy to see. He was a good man, and I have thought of him often since leaving the FAA in 1994 to work at Gulfstream. Now for the "how he changed my life": I have a saying that "chance governs life." Dave wanted me as the candidate for the job and persisted after the FAA's selected candidate did not work out (wrong qualifications, no test

experience). My path to Gulfstream was a result of Dave. The Society of Experimental Test Pilots asked me to host and plan a symposium in Atlanta. Gulfstream's flight test team attended and after the symposium said they were interested in having me to come and work in Savannah. I was hired as a test pilot, eventually became chief test pilot, then director, and then a VP: a total of 23 years at Gulfstream. If not for Dave's persistence, I would not have had that success. For all of life's randomness I realize my good fortune to have crossed paths with Dave. I will miss the fact that Dave is no longer with us. He is in my memory, and I still hear his voice. I will say his name every morning, added to those who have made a difference in my life.

Sincerely,
Randy Gaston"

Paul Sconyers was the assistant office chief at the Atlanta ACO during Dave's tenure. Paul wrote:

"As it always is at times when one loses a loved one or close, dear friend, one never knows what to say. As you know, I worked with Dave for a good many years at the FAA. While we had occasional disagreements over the years, I learned and gained an enormous amount of respect for Dave as a colleague. Many was the time I told Dave when push came to shove, I wanted him in my fox-hole. That means a lot when it comes from a Marine to a Navy man. I'll always look back at the time we worked together with great admiration, but most of all great respect. Dave was a good person, and I will truly miss him."

Bob Sonoba, an engineer who worked on the Phoenix Fanjet with Dave in Lakeland, Florida, wrote to say,

> *"It was a joy and an honor to work with Dave and be his friend."*

Dave Crew (former FAA flight test engineer) relayed a HondaJet memory:

> *"Sad news indeed. He is one that won't be forgotten. Hard to believe, but it was 16 years ago this week that he and Rich Gritter flew the HoJo cert wannabe into Oshkosh for Mr. Michimassa's big coming-out party. Retiree time goes by quickly. My son and I talked with them after they did flybys and taxied in, smiling to the crowd. Mr. Fujino's entourage held an elaborate forum that day to a packed-out tent promising quick FAA certification. Dave and Rich may have been in the background; I don't remember. That was 2005 ... 10 years later, it finally happened."*

* * * * * *

I took some time off from the memoir following Dave's death—for more than a year, revisiting the stories he told was just too painful for me. In the end, I realized the memoir had to be completed—not just for me, not just for the children, and not just for the grandchildren—but for Dave.

—*Becky West*

PROLOGUE

When I finally decided to write a book about my life—after several years of encouragement to do so from my wife, Becky, the biggest question I had was who the audience would be. Should I write this just for family and friends? Should I write it for pilots and anyone else interested in aviation, a field in which I was active for nearly 50 years? Should it be a tell-all book about my time as a test pilot for the Federal Aviation Administration (FAA) and the politics that often influenced decisions about the safety of airplanes and the flying public? Or...should it be just a collection of anecdotes about the interesting things I have done in my life—along with a few of the interesting things that perhaps I should not have done?

In the end, I tried to encompass all the above in the pages that follow. At the core, however, this is simply a book about my life: the things I have done, the people and events that have helped shape me, and some of the adventures and a few of the misadventures I have had. This book will give me the satisfaction of knowing there will be a written record for my family and friends of a life lived almost as much in the air as on the ground. For everyone else who reads it, the book can be whatever they want it to be.

Perhaps some readers will be able to identify with that early portion of my life growing up in a small town on Michigan's Upper Peninsula in the 1940s and 1950s, where I learned about hard work and commitment to family, friends, community, and country that was instilled in me by my father and mother.

Others may find some inspiration in how I was able to overcome a severe injury to my right hand as a child and go on to become a naval aviator and FAA test pilot, despite numerous surgeries over a period of years.

There may be some readers who will identify with the self-doubts I had as a Midshipman at the United States Naval Academy and how I was able to overcome them, primarily through the faith my father had in me.

Still other readers may gain some insight into the early stages of the Vietnam War and the frustrations I felt as the pilot of an F-4 Phantom jet flying off the deck of the aircraft carrier *USS Franklin D. Roosevelt*—mainly due to the lack of clarity about just what we were trying to accomplish in Southeast Asia.

Certainly, there will be some interest in my more than 22 years as a civilian test pilot with the FAA. From 1975 until I retired in 1997 and went into private consulting, I probably had more impact on flying safety in this country than anyone will ever know, or that I will be able to fully explain. Being a test pilot was certainly the most interesting and exciting part of my career.

I have always liked exciting things and flying airplanes—especially testing airplanes—is certainly an exciting way to make a living. That is not to say I was a thrill seeker. You cannot stay alive as a test pilot if you are doing it simply for the thrills. There is an old and well-worn adage among pilots: "There are old pilots and there are bold pilots, but there are no old, bold pilots." I may have occasionally been bold when I was testing a new or re-configured aircraft, but only because I felt it did not endanger the aircraft or me—especially me.

I was so safety-conscious about flying that I once went to a parachute-packing plant in Alabama to see how they packed the 'chutes I wore when I tested aircraft, because I wanted to make sure they were doing their jobs as well as I was expected to do mine.

I was no daredevil, but I was willing to push the envelope as a pilot because I was curious about almost every airplane I flew, to the extent I wanted to see just what its limits were. People have told me they sense something in my personality that makes me want to push the limits in almost everything I do. Maybe there is, but it's not something of which I am conscious. I always believed if I tested aircraft properly and pointed out flaws or problems, the flying public, pilots, and aircraft manufacturers would benefit. If that meant pushing the limits a bit, so be it.

Over the years, I have flown everything from hot air balloons to helicopters, from executive jets to fighter jets, from single-engine passenger aircraft to multi-engine airline jumbo jets. On each of those aircraft, whether flying or testing, I believed there was a right way to do things and everything else was wrong. That was the attitude I took into every test I performed on every aircraft, whether I was flying for the Navy, the FAA, or as a private consultant. There is no compromising on principles when it comes to aircraft: Do the job the right way and everyone gets up and down safely. Do it the *wrong* way, or do not do it *all* the way, and people can die.

During my time with the FAA, I butted heads with management a number of times over testing of new aircraft and new systems. I had a reputation of being argumentative, a contrarian, someone who did not "go along to get along" with either the bureaucracy or the airline industry, the latter which continues to have a great deal of political and economic influence over the government agency entrusted to oversee the industry and the safety of the flying public.

One issue on which I tangled with my bosses all the way from the Atlanta office to the FAA headquarters in Washington, D.C., that I will go into in more detail later in the book, occurred in the 1980s and involved what was then a new collision avoidance system for passenger aircraft. When I questioned some shortcomings in the system, I was told to shut up and go away. The response I got from my bosses was, "Well, we told Congress we would have the system certified by such and such a date, and you're not going to slow that down."

It seemed the FAA hierarchy cared more about meeting a deadline than it did about whether the system worked as intended. I was concerned that they were in too much of a rush to get a flawed system into airplanes that could end up killing people. When I raised my concerns, I was told, "Oh, that's not going to happen."

Efforts by my superiors to shut me up and make me go away were unsuccessful. If anything, those efforts only encouraged me to speak more openly about the problem. Eventually, the changes for which I and a co-worker were pressing were implemented and became part of the standard equipment on all passenger aircraft.

I saw that "go along to get along" mindset within the FAA and the airline industry many times during my years with the agency. There was a cozy relationship between the FAA and the aircraft manufacturers, especially the larger companies like Boeing, Gulfstream, Piper, Beech, and Airbus. Those companies have a lot of money, and that money is able to buy plenty of political influence. If the FAA made a decision that one of those companies did not like—especially if they thought it would cost too much money—executives would not argue with the FAA; instead, they complained to the member of Congress representing their district and the pressure would come from the politicians.

Eventually, it got to the point that the FAA, because of a constant lack of qualified inspectors and pilots, allowed the industry giants to take more responsibility for certifying new systems and new aircraft. It was like giving the aircraft manufacturers *carte blanche* to cut corners when they thought it would save a few dollars and shorten the certification process.

My frequent battles within the FAA notwithstanding, I thoroughly enjoyed my time as a test pilot for the agency. Usually, I was responsible only for the aircraft and myself; I did not have to worry about meeting a particular flight schedule or trying to make grumpy passengers less grumpy. Those are just two of the reasons I never had a real interest in flying for the airlines, although I did receive an offer from a major airline.

Sure, airline pilots make a lot of money, but, to me, it was like being a bus driver: Instead of driving down the highway, you're driving down the airway. Over the years, I have flown too much on too many different airlines with too many cranky passengers to know that I would not have been Captain Cheerful in the cockpit of a commercial airliner. I had it in my mind that flying commercial aircraft would be too confining, and I wanted to do something that enabled me to be a bit more independent than I would have been with an airline.

I guess I have always been independent in that way, setting a goal and striving to reach it, even though there might be more tantalizing opportunities around. I seemed to enjoy reaching the goal, no matter the obstacles, and not allowing myself to become absorbed in those other opportunities, no matter how lucrative they seemed at the time.

Other people apparently saw that focus in me, even if I did

not see it in myself. While I was still working for the FAA, I once interviewed for a Senior Executive Service (SES) position. Several people, including an assistant division chief from the FAA's Dallas office, conducted the interview in what was a role-playing exercise. After so many years, I do not recall exactly what the exercise consisted of, but when it was over the head of the panel briefed me.

"Dave," he said, "I have never seen anybody as goal-oriented as you. I couldn't get you off the track, no way. You knew where you were going and how you were going to get there."

I honestly do not see myself as being goal-oriented to the extent that it impedes all other thought, which it can with some people. I was just blessed with the ability to handle a number of things at the same time while keeping my focus on whatever long-term goal I had at the time. It was that way with flying and began more than a decade before I ever took my first flight. And, to a certain degree, it has been that way with this book. Once Becky convinced me to write my memoirs, it became a long-term goal I wanted to accomplish before I got too old to remember some of the things I did so many years ago.

So, as you read this book, take from it what you will. If you get some insight into my love of flying—or get a few chuckles from some of the stories I tell—so much the better. Just remember: This is my story, and I am sticking to it, no matter what anybody else says.

CHAPTER 1
GROWING UP A "YOOPER"

The plane may have been a DC-3 or something a bit smaller. I was too young to know exactly what I was looking at as the plane flew over our house in Ironwood, Michigan, that warm summer evening in the early 1940s. I was only six years old, or maybe seven, at the time, but that low grumble of engines and the sun glinting off the silver hull stirred something in me and planted a seed in my mind that someday I would be a pilot.

I am not sure how or why that happened. There was nothing in my family's history to suggest that flying airplanes would become my vocation and passion. To the best of my knowledge, the only person in my family who had ever flown was my Uncle Toby, who served in the U. S. Army Air Corps during World War II. Certainly, my parents had never flown. Flying was expensive, and if my parents wanted to go somewhere, they walked, took a bus or train, or caught a ride with somebody who had a car. At that time, my parents did not even own a car.

As I watched the plane fly over, I turned to my mother sitting next to me and said, "Mom, one of these days I'm going to fly one of those." I'm sure she smiled and likely thought it nothing more than a childhood fantasy, as mothers often do when one of their children makes a prediction that seems so outlandish that it was not unlike saying: "Mom, one of these days I'm going to be President of the United States."

That was not the first time my mother and I watched airplanes from our back steps. In the summer after dinner, we often went out there to enjoy the warm evening air. The steps faced north toward what is now known as the Gogebic-Iron County Airport and when planes took off, they often came right over the house. On this particular evening, though, I set off on my life's path almost without realizing it.

The idea of my flying an airplane was not a goal that was all -consuming, as it was with some pilots I later came to know. I did not build model airplanes and hang them from the ceiling of my room. I did not put pictures of airplanes on the walls, as that was something my father never would have permitted. Nor did I read much about aviation as a youngster. Still, the goal was there, and I continued to work toward it even though it was more than 10 years before I took my first flight and another five years after that before I became a pilot.

*****

My birth on Sunday, November 13, 1938, was decidedly unspectacular, except perhaps for my mother, although I do not recall her mentioning anything to me about problems I may have given her that day. I was the third child and only son born to Holger Frans West and Helen Heczko West of Ironwood, a thriving iron ore mining and logging community on Michigan's Upper Peninsula just south of Lake Superior.

My older sister, Ruth, was four at the time. The middle child, Margaret, was two. By the time I arrived, I am sure both my mother and father knew what to expect and how to deal with things as parents. And, of course, my sisters were there to help me quite a bit during my early years.

The next day's edition of the *Ironwood Daily Globe* made no mention of my birth. The front page that day was filled with headlines of events that hinted at the coming conflict in Europe that was about to turn the world upside down while it also told of more mundane happenings of local interest.

A story from Berlin carried the headline "Jews Banned From Schools." Another from Pittsburgh told readers "Oppression Of Jews Assailed." A story from Los Angeles under the headline, "Legally Dead Man Brought to Life," told of a man who faked his own death only to be caught years later living under an assumed name.

Locally, readers were informed of a "Snow-Less Day Prospect for Deer Season Opening." And "Mrs. R. Douglas Is Hurt in Fall," which told of an Ironwood woman who was so frightened by fireworks going off outside her house that she fell and broke her hip.

Ironwood was the essence of small-town pre-World War II America, and my parents were the essence of first-generation Americans, both the children of immigrants. They were quiet, hardworking, and devoted to family, community, and country.

My father's parents, Matthew and Elvira West, claimed to be Swedish, but for reasons never explained to me, Grandfather West said he had been reared in Finland before he came to the States in 1902 at age 12. My mother's parents, George and Susannah Heczko (pronounced HETCH-ko) were from an area of Eastern Europe known as Silesia. The Heczkos used to say that area had been fought over so many times they were not sure if they were

German, or Polish, or Czech. Although both spoke Polish almost exclusively while I was growing up, they were Lutherans, a predominantly German religion.

Grandmother Heczko called me "Davedek" (pronounced DAH-vah-deck), which I learned only much later is most likely the Polish equivalent of "Davey." I think I was one of her favorite grandchildren because I spent so much time at the farm she and my grandfather owned outside Ironwood. Every so often, she would slip me a $10 bill when no one was watching, which was a lot of money to a kid in the 1940s.

Grandfather Heczko came to the United States by ship in 1908. Grandmother Heczko and the two children they had at the time—Anna and John—came in 1910. Grandfather Heczko was supposed to get off the ship in New York City. However, because of his inability to understand English, he did not realize the ship had arrived in New York and he stayed aboard until it reached Houston, Texas. Eventually, he found his way to Michigan's Upper Peninsula, or the UP, as most people in that part of the world call it. Those of us who lived there were known as UPers, pronounced "Yoopers." It was a title we wore proudly wherever we went, if for no other reason than we considered ourselves tougher than those other Michiganders who lived "down south" on the other peninsula.

In the old country, Grandfather Heczko had been an indentured servant working the land for its owner. One of his jobs was to scare up game for the nobles when they went hunting, but he was not content to spend his life working for someone else. He wanted his own land and his own farm, but he knew he had to leave Europe to achieve this goal. By the time he died in 1953, he owned nearly 200 acres of prime Michigan farmland and considered himself a wealthy man.

While the Heczko farm was about seven miles south of our

house in Ironwood, my father's parents lived just a couple of blocks from us. Grandfather West had his own piece of property, about three or four acres, but that was enough for him. On the east side of that property, my father built a small woodworking shop where he often went after dinner to work on a project or two. Sometimes I went with him to sweep out the shop or to help clean up the odd bits of wood lying around.

Ironwood was in the heart of what in the 1930s and 1940s were thriving iron ore and lumber industries. The mines were the economic lifeblood of the area, and that was especially true of Ironwood. My father was not a miner, though. He was a carpenter who worked at the Geneva Mine in Ironwood, one of eight iron ore mines in the immediate area. Each mine employed dozens of men who brought home decent salaries for those days. They were tough, rugged men whose work in those mines, some of which were more than 2,000 feet deep, was dark, dirty, and dangerous, not unlike coal mining.

Although my father called himself a cabinetmaker, he was much more than that: He was a master craftsman when it came to working with wood. When he wasn't working at the mine, he built and remodeled houses, built cabinets and bookcases, and repaired just about anything made of wood. He even built our house the year I was born.

He purchased two lots in the Mayview Subdivision in what was known as the Aurora Location in Ironwood for $150 in February 1938. By the time I was born in November, the two-story, three-bedroom house at 922 East Pine Street was finished, and that's where I grew up. Our house was at the end of a dead-end street with a vacant lot next door. The land in that lot was not usable for anything and there was a big rock on the lot where I spent a lot of time playing as a youngster.

5

My father had a reputation for knowing his woods better than anyone around. One day one of the workers came up to him, handed him a piece of wood and asked, "Holger, what kind of wood is this." My dad held it up, looked at it, smelled it, and said, "That's piss elm." Then he tossed the wood back to the worker. The man had urinated on a piece of elm to see if he could fool my father; obviously, he could not.

Summers in the UP are a long time coming and a short time staying. Whenever winter decided to leave us and the ground thawed enough for planting, Grandfather Heczko was out in the fields working. I spent a lot of time on the farm and enjoyed the hard work, even as a youngster, helping whenever and however I could. But, that eagerness to help on the Heczko farm one summer nearly destroyed whatever dreams I might have had of becoming a pilot.

During the haying season of 1943 when I was not quite five years old, Grandfather Heczko and some other men, including my father, were gathering hay from one of the fields to be stored in the barn loft until winter, when it would be fed to the cattle. They brought the hay to the barn on a horse-drawn wagon. Then, large tongs were lowered by a rope to pick up the loose hay from the wagon bed and lift it into the loft.

The rope used to lift the tongs and the hay was strung through a series of large, wooden pulleys that ran from the loft across the upper portion of the barn and down one corner on the outside. One end of the rope was attached to a team of horses that was led away from the barn once the tongs had latched onto the hay, thereby lifting the hay into the loft. Once the hay was in place, the horses were backed up, so the rope slid back through the pulleys, lowering the tongs for another load.

On that day, I decided I was going to help the horses pull the rope. I grabbed onto it and before I realized what was happening, my hand was pulled into one of the pulleys and was trapped there. As the rope slid through the pulley, it began shredding the palm of my hand.

I do not remember crying or screaming or feeling any pain at the time. I am sure some of the shock of the experience camouflaged much of the pain. However, I remember there was a great deal of blood, and my first thought was that I needed to wash the blood off my hand.

I walked into the kitchen of the farmhouse and asked my aunt Gertie Heczko, who was standing at the sink, for a bucket of warm water. She did not see my hand, but she gave me the bucket of water and I went to the back porch. When she came out and saw the blood and my shredded hand, she sucked in her breath and exclaimed, "Oh, my goodness!"

As soon as my father learned what had happened, he rushed in from the field, looked at my hand, and called Dr. H.A. Pinkerton, who ran the medical clinic at the Geneva Mine.

"Dave's been hurt!" said my father, quickly explaining what had happened. "We're bringing him in now. I want you to look at him."

"We're not open," Pinkerton told him. It was Sunday, and the mine and the clinic were closed.

My father was not a man who got angry easily, but he told Pinkerton quite bluntly, "When I get there, if you're not there, I'm going to have the police pick you up and bring you there."

Not surprisingly, Pinkerton agreed to meet us at the clinic. When he arrived, Pinkerton looked at my hand, put some sulfa on the wound, and bandaged it. Then, he told my father, "We can fix that in two or three surgeries."

My parents had no idea how severe my injury was, or what sort of reconstructive surgery I might need. In those days, if a doctor said he could do something, people usually believed him. Few people thought about second opinions. And I suppose they believed that if Pinkerton did not have the expertise or experience to do it, he knew someone who did.

I never found out exactly what Pinkerton was suggesting in terms of surgery, but during my year in kindergarten, the school nurse visited our house to check on me. Elizabeth Phelan had served as an Army nurse during World War I and had seen a number of traumatic injuries worse than mine. She examined my hand and saw that the pulley had taken virtually all the flesh off the palm of my hand and had severely damaged several of my fingers. When Elizabeth Phelan heard what my parents said Pinkerton had recommended, she shook her head.

"Don't let him do that," she urged. "I was an Army nurse and I've seen a lot of things done in restoration." She recommended reconstructive surgery on my hand and suggested my parents contact what was then known as the Michigan Society for Crippled Children.

In the summer of 1944, I was taken to the Northern Michigan Children's Clinic at St. Luke's Hospital in Marquette. On the day of the first surgery, I was wheeled into the operating room and administered ether. I was fighting the anesthesia and vividly recall the nurse saying urgently, "Doctor, his temperature is up to 105! We need to do something quickly! If we don't get it down, we may lose him!"

When I woke up, I was in a strait jacket, my right hand sewn to my abdomen in what was then called a "full flap procedure." This allowed flesh to grow back on my hand from my abdomen. It was a more complicated procedure than a skin graft. A skin graft would not have allowed the skin to grow back because the blood supply to my hand had been damaged and a graft would not have taken.

8

Years later, when I hit puberty, I noticed I was growing hair on my right hand because that was skin that had come from my abdomen. To this day I have to shave my right hand occasionally so I don't have a hairy palm.

For the next three months, my entire summer vacation that year, I lay in that hospital bed in Marquette, more than 145 miles from home and family. The only time I saw my parents was when one of my uncles, Paul Heczko, was able to drive them from Ironwood to Marquette for a Sunday visit.

That single visit was brief, a few hours at most, before my parents had to get back in the car and return to Ironwood in time for work the next day. They were there and then they were gone again. It was almost worse than not seeing them at all.

That summer was probably the longest of my life. I have tried to block most of it from my memory, but I still remember how slowly time moved as I lay in that bed staring at the ceiling and the walls around me. There were four or five other kids in the ward, but we had nothing to do: There was no radio, no television, and no visits except occasionally from a nurse or a doctor.

The orthopedic surgeon who performed the initial surgery was Dr. Eugene Elzinga. He was kind enough to stop by my bed occasionally to see how I was doing and joke with me to lift my spirits. I guess he felt sorry for me because I was so far from home and was not receiving many visitors.

One day as we were chatting, I asked Dr. Elzinga where he got the stitches he used to attach my hand to my abdomen. He smiled and said, "My daughter Judith has some horses and I go out and cut some of the hair from their tails, and that's what I use for stitches!"

I wasn't sure whether to believe him or not, but I appreciated the fact that he thought enough of me to stop by every so often and tried to make me smile. Being stuck in a hospital bed with nothing to do for that long was a tough row to hoe, especially at my age.

That surgery was the first of 21 I had on my hand over the next 15 years. They did not end until 1959, when I was in my second year at the U.S. Naval Academy. Several of the early operations were at the hospital in Marquette. Several after that were at Blodgett Memorial in Grand Rapids, Michigan, and after that, I went to the University of Michigan in Ann Arbor. As the surgeries progressed, I regained more use of my hand, and the time between surgeries stretched from every summer for the first three years to every four or five years.

Each surgical visit meant spending about two weeks in the hospital. As the years went by, the surgeries became more cosmetic than reconstructive, shaping the hand so that it looked relatively normal. Every doctor who has worked on my hand has been surprised that I still have feeling in it. I am not sure where I would be today had my parents not had the foresight to take the advice of Nurse Phelan over that of Dr. Pinkerton. I probably would have spent my working life as a $3.50-an-hour left-handed bricklayer.

When I returned to school in the fall of 1944, the Michigan Society for Crippled Children placed me in a school for children with special needs. There was a single classroom for everyone from 2nd to 8th grades. Some of the kids in that class had polio and some had other special needs, but I did not feel that I fit into any of the special needs categories and was not sure why I was sent there. However, it helped me academically so that by the time I left that school at the beginning of my 6th-grade year and returned to a regular classroom, I was far ahead of my peers, surprising my teachers. It was probably the more individualized attention that I received that gave me that edge.

Otherwise, I was never limited in what I could do. I had several part-time jobs throughout childhood and into my teenage years. I worked at two theaters as an usher. I worked as a clerk at JCPenney's. I set pins at the local bowling alley. And, I had several newspaper jobs at different times, delivering the *Ironwood Daily Globe*, the *Milwaukee Journal*, or a weekly newspaper known as *The Grit*, which billed itself as "America's Greatest Family Newspaper" with an audience primarily in small towns and rural areas.

I also started playing football when I was in the 6th grade and later played linebacker and fullback for the L. L. Wright High School Red Devils. Our nickname came from the red dust that coated the faces and hands of the miners after a day digging iron ore in one of the local mines. In 1954, when I was a sophomore and a little-used substitute for the Red Devils, we won the Class B state title. I was a starter my junior and senior years, but we were a bit less successful those years. My surgically repaired hand was never an issue.

In addition to all those activities, I worked on the Heczko farm every summer, partly because I enjoyed it and partly because as a family member, I was cheap labor: Haying season started in June. By July and August, it was time to harvest grain. If we were not harvesting hay or grain or tending to the dairy cattle, we were gardening, digging potatoes, picking vegetables, or occasionally slaughtering a pig. I had several cousins who lived nearby but they tended to shy away from farm work, maybe because it's usually hard work. I was one of the few Heczko grandchildren who volunteered to work on the farm on a regular basis.

One of my least favorite farm jobs as a youngster was to hold a large bucket to catch the blood when we slaughtered a pig. Uncle Paul had been in the Marine Corps and brought home a service knife that he used to kill the pig. The pig was laid out on the ground, Paul stuck it with his knife, and I would catch the blood in the bucket.

We usually did the slaughtering in the late fall when there was snow on the ground. We filled buckets with just a little snow to keep the blood from coagulating so it later could be used to make blood sausage. Known in Polish as *kiska* or *kishka*, blood sausage is just what the name implies: a sausage casing filled with blood, spices, some sort of grain, and then pan-fried like any other sausage.

We used virtually every bit of the pig after it was killed. Grandmother Heczko used to say the only thing we threw away was the squeal. Besides the usual ham, bacon, and pigs' feet, one of the other delicacies was something called head cheese. This is not cheese in the traditional sense of it being a dairy product; rather, it is more like a meat jelly that comes from the flesh of the pig's head. It's seasoned, cooled, and then cut into slices. Grandfather Heczko used to eat head cheese sandwiches for lunch, often after a breakfast of pig brains and scrambled eggs, one of his favorite meals.

Both of my grandfathers were quiet men. Grandfather West, who worked in the mines, never had much to say and did not interact with his grandchildren a great deal. I do not recall spending much time with him although he sometimes asked me to come over and cut the clover in his garden.

Grandfather Heczko had a limited command of English, but he never talked much anyway. I think he understood more than he spoke, but that's just my guess. He rarely got upset unless someone was bothering his cows. Those cows were almost like his children. He could yell at them, swat them, kick them, or do whatever he wanted; but let somebody else mistreat them, and there was hell to pay. One Sunday several of my younger cousins were at the farm with me and they started throwing rocks at the cows. I don't think

I ever saw Grandfather Heczko as angry with his grandchildren as he was that day, chasing them away from the cows with some choice words in Polish.

Spending time on the farm was always enjoyable. Since my parents did not have a car, we usually walked the seven miles from our house in Ironwood to the farm. Often, my mother would bring Grandfather Heczko several bottles of beer, and that always made him smile.

When I went to the farm early in the mornings, I often was greeted by the smell of Grandmother's freshly baked *kolaches*. She usually got up at 4 a.m. to make them fresh in her wood-burning stove, topping the circular pieces of puff pastry with fresh blueberries, blackberries, or apples. Yet, no matter how tasty the *kolaches* were, and no matter how much I enjoyed farm work, neither was enjoyable enough to convince me to become a full-time farmer. A gentleman farmer, maybe, but working on a small farm eking out a living was not one of my long-term goals.

My mother also was a wonderful cook, a skill I am sure she acquired from her mother. One of my mother's specialties was pasties. These are meat pies, similar to an empanada. The word is correctly pronounced PAST-ee, not PAYS-tee, the latter which are something modest strippers wear. Many of the men who came to work in the iron ore mines in the Upper Peninsula were of Welsh descent, where pasties are a dietary staple. Pasties easily fit into a lunch bucket and could be eaten with one hand, like a hamburger, so they were convenient for the mineworkers.

My mother could cook just about anything well, though. She used to joke that when a cow stopped giving milk, it became beef and sometimes that beef was a little tough. But my mother had a way of cooking it so that no matter how tough it was, it tasted pretty dang good.

Not that anybody complained much about what we had to eat. Nobody in our house was particularly fussy. Besides, if you were, it didn't get you anywhere with my father. If one of us said to my mother, "I don't particularly like that," my father usually replied, "Okay, you're excused, but don't come back until breakfast." It was either to eat what was offered or to go hungry all night.

When I was about 12 years old, I started raising rabbits to sell for meat. A man I knew lived next to the sawmill in Ironwood and raised rabbits and pigeons. I used to go down there to pet the rabbits and learned that a number of people enjoyed the taste of rabbit. I asked my dad if I could put some rabbit cages in a storeroom on the side of the house where he kept tools so I could raise some rabbits to sell. He said I could as long as I kept it clean, which meant getting rid of all the rabbit manure on a regular basis.

I started with one female rabbit I named Patrina. She quickly had a litter and before long I had six adult female rabbits and one or two adult male rabbits. Six female rabbits can produce a bunch of bunnies. I raised the rabbits until they were several months old and then sold them. I had a pretty good business going between friends and family and neighbors wanting rabbit meat.

The only trouble was I needed some help killing them before I delivered them to my customers. My mother, even though she was a farm girl, had more than a little trouble with that aspect. I held the rabbit by the back legs while she smacked them over the head to stun them, and then I cut off their heads. I could tell my mother did not particularly enjoy that part of it, but we worked through it and made a few dollars selling rabbit meat. We also had plenty of rabbit manure for the vegetable garden. I ran that rabbit business until I graduated from high school and went away to prep school prior to heading off to the Naval Academy.

I also raised ducks one year. I won some ducks at a carnival and brought them home. When they got large enough, we decided to have one for Thanksgiving dinner that year. Unfortunately, I had fed them so much grain they were greasy as heck, and we could hardly eat them. I later learned that what you need to do with ducks and geese is parboil them first to get rid of most of the fat, and then you roast them.

I also hunted for anything we could eat, especially wild rabbits, ducks, geese, partridge, and deer. My father was not particularly fond of hunting, so I usually went with my Uncle Paul. When I was about 13, I wanted to buy a 16-gauge Ithaca pump shotgun. It cost about $100, which was a considerable sum of money back then. Although my father was a skilled laborer, money was always tight for us and if I wanted something, I had to earn the money to buy it. My father worked out a deal with the owner of the hardware store so if I paid $5 a week, I could have the gun while I was paying it off. It took several months, but I ended up buying that shotgun with my own money. My younger son, Travis, has it today.

One thing my father loved was fishing. We lived only about 15 miles from Lake Superior and in the summer he, Uncle Paul, and I would go fishing for lake trout. One day in 1947, my father hooked onto something big. It turned out to be a rainbow trout, about 28-30 inches long, which is a good-sized trout. On the way home, we stopped at a nearby bar. It was summer, and summer brought crowds of vacationers from Milwaukee and Chicago to go fishing or camping around the lake. Whenever you went into one of these bars near the lake and people realized you'd been fishing, they always wanted to know if you'd caught anything.

"Yes, I caught one rainbow trout," my father said.

"Do you mind if I have a look at it?" one man asked.

My father agreed and we went outside to pull the fish out of the cooler to show it off.

When the man looked at the fish, his eyes got big, and he said, "I'll give you $25 for that fish."

We certainly could have used the money, but before responding to the man, my father turned to me and said, "Dave, do you think we ought to sell it?"

I thought for a moment and then said, "Nope, we're going to eat him."

My father turned back to the man and said, "It's not for sale."

I think I realized at the time that the man probably would have taken that trout back to Milwaukee or Chicago, had it stuffed and mounted, and then bragged to everyone about *his* great catch.

That particular moment had a much bigger meaning for me, though. It told me that my father had a great deal of faith and respect in my judgment, even at my young age. It is a moment I cherish to this day.

Summers on the UP can be pleasant, except for the mosquitos that infest the region because of all the nearby water. Winter is another story. We used to joke that the UP has two seasons: winter and more winter. Actually, it's not a joke. If you don't like snow, don't go to the UP in winter. About 120 miles northeast of Ironwood, on the Keweenaw Peninsula jutting northward into Lake Superior, is the town of Calumet. One of Calumet's attractions is the "Snow Thermometer." The record of 390.4 inches was set in the winter of 1978-79; that's about 32.5 feet of snow in one winter.

That's typical of the UP, though. The January before I was born, Ironwood had a storm that dumped more than 30 inches of snow in about 36 hours. It was accompanied by 50-mile-per-hour winds that created drifts that nearly covered telephone poles. Nothing moved for three days, as the technology to clear roads and de-ice telephone and electrical lines was not nearly as sophisticated or efficient as it is today. Over Easter weekend, when I was a junior in high school, we got 40 inches of snow.

So what do Yoopers do when winter comes and doing anything outdoors is usually more trouble than it's worth? Most of them find ways to entertain themselves indoors in the evenings. For my family, in an era without television, that meant playing cards. My father loved to play cribbage. If you don't know what cribbage is, or how to play it, it is far too technical to explain here. Suffice it to say that it involves numbers and counting. My father taught me to play cribbage when I was about four. I also learned to play smear, canasta, pinochle, poker, and bridge over the years.

On those long, dark winter nights after dinner, my father often asked, "Do you want to play some cards, Dave?" And I would say, "Sure," and out came his prized cribbage board. I still have that cribbage board; it has become a prized family heirloom. It always reminds me of my father and those nights we spent playing cards together.

I learned two things from playing cards. The games taught me to be competitive at an early age, and I'm still extremely competitive when I play cards. When I play, there is no talking about anything but cards. I never lost deliberately, even to my kids when I taught them how to play years later; they knew that and understood that if they were going to beat me, they had to beat me fair and square.

I also played canasta with my Grandmother West and a few ladies she invited to her house on occasion. When I was eight or nine years old, she sometimes called down to our house and said, "David, would you like to play cards today. And, I would say, "Sure" and run up there and play cards with these little old ladies. Usually, I won.

Cards later became a wonderful way to pass the time at the

Naval Academy, where we covered the windows with blankets after Lights-Out and spent half the night playing pinochle. When I was aboard ship during my deployment to the South China Sea years later, pinochle was the favored game, and with my experience at the game I was able to make a few extra dollars when we played for money.

The other thing I learned from playing cards is how to add and subtract at an unusually early age. My mother loved to tell the story about having a parent-teacher conference when I was in kindergarten and the teacher telling her, "Dave can count, and he knows his numbers so well. Where did he learn that?"

My mother smiled sheepishly and said, "I probably shouldn't tell you, but he and his father play cards together, so he's gotten to know his numbers pretty well."

Growing up in Ironwood was a little like growing up in Andy Griffith's Mayberry. It was safe, peaceful, and as wholesome a place to raise a family as anyone could want. However, just across the Montreal River, less than a mile from our house was Hurley, Wisconsin, a place many people considered the Sodom and Gomorrah of our little part of the world. Crossing the river on the Silver Street Bridge from Ironwood to Hurley was like stepping into another world.

Hurley may have been the toughest and wildest town in the region. A 1941 U.S. government publication titled *Michigan: A Guide to the Wolverine State* was quite complimentary of Ironwood but had little good to say about its next-door neighbor to the west:

"Throughout the Middle West, wherever lumberjacks and miners congregated, Hurley was known as the hell-hole of the range," the guide read. Penned by writers in the Federal Works Project Administration during the Depression, the guide pointedly noted "the sin, suffering, and saloons that gave Hurley a reputation unrivaled from Detroit to Duluth." A 1954 book titled *Hurley: Still No Angel* by Lewis C. Reimann, provided a complete list of Hurley's 63 bars and taverns (down significantly from the 110 or so during Prohibition).

I was convinced the Mafia ran Hurley. During Prohibition, the likes of Al Capone and John Dillinger came to the area to get out of Chicago during the summers. One of Capone's brothers, Ralph, died in Hurley in 1974.

Once I got to be a junior and senior in high school, it was not unusual for me and my best friend, Ed Shiffra, to go to Hurley. I have known Ed since we were in kindergarten and shared a blanket during nap time. You only had to be 18 to go into a beer bar in Wisconsin at the time, but in Hurley, the liquor laws often were ignored if you had the money. Besides, Ed and I had people in Hurley who looked out for us whenever we went over there, and they likely were tougher than the Mafia—they were Ed's mother and father and one of his uncles.

Ed's uncle owned one of the clubs, in Hurley, a place called Club Carnival. It's now a strip joint, but back in the early 1950s it had a good restaurant and bar and was considered one of the more respectable places in town. Ed's father was a bouncer at the club, and his mother was a bartender. One night Ed and I were in there and these hookers came in and started flirting with us; they were trying to get us to partake. Ed's mother saw what was going on, hurried over to where we were sitting, and said, "You girls get out of here! This is my son and his friend! Leave them alone!"

Although I was a bit of a hell-raiser when I got into my teenage years, I knew that if I did anything too outrageous, I not only risked incurring the wrath of my father, but I risked throwing away my dream of becoming a pilot. Probably the closest I came to getting into any kind of real trouble was at Zinzarella's Pool Hall on

Aurora Street in Ironwood. The place was run by William Zinzarella and his wife, Pia. He was from Austria, and she was from Italy, and both were old fuddy-duddys who did not like kids coming around to play pool. Mrs. Zinzarella wore some high-top shoes that looked like combat boots, which always amused us.

The Zinzarellas got really irritated when one of us bounced the cue ball off the table and it rolled under the table. That would set off Mr. Zinzarella. He would come over raising hell and when he tried to retrieve the ball from under the table, we gathered around and wouldn't let him out. Then Mrs. Zinzarella would call the police. The police would come and tell us to break it up and go home. We did but usually came back a week or so later and did it again. The police probably spent more time in Zinzarellas than any other place in town, but none of us was ever arrested.

By my last two years of high school, I was locked in on going to one of the service academies to learn how to fly. I briefly toyed with the idea of becoming a lawyer, or a doctor—several people told me they thought I would make a good lawyer—but neither of those professions seemed to be a very exciting way of making a living. Maybe that's why I focused on flying; I could see something special in flying and wanted to do something exciting. I certainly did not want to go down in the mines.

For many people in that area of the Midwest, the mines were a way of life, and the sons of miners were expected to become miners. It was a booming industry in the 1930s and 1940s and the money was good for those who worked in the mines. Grandfather West worked in the mines and thought my father should work there as well. When my father expressed an interest in going to college, Grandfather West told him, "You don't have to go to college. The mines pay you good money. You get a good job at the mine, and you'll have all the money you need."

As long as the mines were going full bore, there was some truth to that observation. I don't think my father was ever truly happy working at the mine — even topside. He had to belong to the union there, and he was not fond of unions. Every couple of years the miners went on strike, angering my father. Most of the strikes were resolved peacefully because most of the people who worked at that mine lived in that community and knew each other. Their wives shopped together, their kids went to school together, they went to church together, and socialized together, so the strikes were more a formality than something that lasted long or became too divisive. Eventually, though, my father left the mine and the union and went to work as a superintendent in a plant that made travel trailers.

I think somewhere along the line my father decided that if he ever had a son he would do whatever needed to be done to make sure that son went to college. He bluntly told me that he resented his father for not helping him go to college, so he wanted me to have the opportunity he never had. He knew I did not want to be a miner, and I knew there was no way I was going to be a pilot if I stayed in Ironwood. I liked Ironwood; I had a good time there, and it was a good place to grow up. But, it was a one-business town, and by the 1950s, the end of the iron ore boom was on the horizon. Not too many of my classmates stayed in Ironwood after graduation. Many went to Milwaukee, Chicago, or southern Wisconsin. When the mining jobs started going away, so did the young people.

By 1960, four years after I graduated high school, my father also saw the economic hard times coming for Ironwood. He and my mother left for Rhode Island, where my sister, Margaret, and her family lived. He worked there as a carpenter until he died in 1970 at the age of 57 in a fall from an icy scaffold. My mother worked as a librarian at the University of Rhode Island in her later years.

Throughout high school, I was building my résumé through a variety of activities I was told would make me a more attractive candidate to one of the service academies. I played football for four years. I was in the drama club for four years, even performing in school plays, and I was sports editor of the school yearbook, the *Hematite* (named for the primary ore in iron). I was in the National Honor Athletic Society, and I was class vice president for three years. I even joined the Michigan Army National Guard.

My reason for joining the Guard was more mercenary than practical—the state sent me $30 every quarter just for showing up for drill once a month. I was Private First Class David P. West of Battery A, 300th Anti-Aircraft Artillery Battalion (75 millimeter guns). I was not actually a gunner, though; I was assigned to the Communications Section.

In retrospect, perhaps one of the most important things I did that helped ensure my eventual appointment to one of the service academies was to date the mayor's daughter. (I was politically connected before I realized it was a good thing to be politically connected!) Gene Velin was the mayor of Ironwood, and I dated his daughter, Barbara, for a while. The mayor liked me, as did his brother, Conrad, who owned a large grocery store in town. Conrad Velin was good friends with our local Congressman, John B. Bennett, and almost without realizing it, I was forging relationships that were beneficial to my future.

In those days, each member of Congress could have as many as five appointments at each service academy at one time—one in each class with a fifth person in any one of the classes. These days, if the member of Congress has three or four prospects he believes are good candidates, he submits their names to the service academies, which then choose the ones they want.

I wrote a letter to Bennett requesting an appointment to one of the service academies, and I received what at that time was re-ferred to as a second alternate's appointment, which meant I was on the list of candidates, but was not necessarily guaranteed a spot. Because the service academies thought I had potential, it was suggested that if I spent a year at prep school after high school to enhance my academic credentials, I would have a better chance of getting the appointment I wanted. My parents had to pay for that year, but it demonstrated to me again just how much my father wanted me to go to college.

I graduated high school in June 1956 and spent the summer working on the Heczko farm. In September, I headed off to Northwestern Prep School in Minneapolis. The course work was heavy in math, sciences, and English. I was always fairly good in math and science, but English and foreign languages were a struggle for me. At prep school, there was little time for much but classes and study, so I was able to focus on what I needed to do to be able to move on to one of the academies.

In February 1957, I traveled to Washington, D.C., with George Mathes, whom I met at Northwestern and later went with me to Annapolis. We went there to meet our respective members of Congress and lobby in person for appointments. I was able to use my status as a member of the Michigan Army National Guard to get a military flight there. Whether that trip was the deciding factor or not, I'll never know. But, that spring, I received a letter from Bennett informing me I was being given a primary appointment to the U.S. Military Academy at West Point.

A few weeks later, I received another letter informing me that I also had an appointment to the U.S. Naval Academy. I might have received an appointment to the U.S. Air Force Academy in Colorado Springs, Colorado, but after the Air Force personnel examined my hand, I was told I was medically unable to perform — even though I had no physical limitations. The Navy and the Army gave me a five percent disability for my hand but said that otherwise I was good to go. Perhaps the answer is that the Air Force Academy was still rather new and had far more restrictive criteria at that time than the other two academies.

After receiving the appointments and pondering which to accept, I went back to Ironwood and discussed my options with some of the people I had worked with at JCPenney. One of the sales associates, a Navy veteran with whom I worked closely, told me, "You'll be making a mistake if you don't take the Navy appointment."

He had earlier helped fuel my desire to fly carrier-based jets by telling me stories about flight operations on a carrier and how exciting it was to be a part of it. Besides, it just looked like the most fun thing to do, and they paid you to do it. And, to the best of my knowledge, the Army does not have many aircraft carriers.

My father obviously was pleased that not only was I going to college, but I also was going to one of the best universities in the country, the U.S. Naval Academy. Not all my relatives were as pleased with my choice, though. When I told Grandmother Heczko I was going to the Naval Academy, she cried. "I don't want you to go. They kill people." The wars in Europe when she was a child were still fresh in her memory.

In an effort to convince me not to go, she said she would give me the farm that Grandfather Heczko had spent so many years nurturing. I tried to let her down as easily as possible and told her that flying airplanes was something I felt I was destined to do. She reluctantly accepted my decision.

On July 1, 1957, I was discharged from the Michigan Army National Guard so I could be sworn in as a Midshipman at the U.S. Naval Academy. That same day I flew for the first time as a passenger on a commercial plane from Minneapolis to Washington, D.C., on my way to Annapolis, Maryland.

I do not remember much about that flight or the long taxi ride from the airport to Annapolis. However, when I got to the main gate of the academy and saw the imposing structure of Bancroft Hall, which was to be my home for the next four years, I knew I had stepped into a completely different world than the one from which I had come.

What I did not realize at the time was just how much of a challenge that new world was going to be. Nor did I realize what a key role my father eventually would play in helping me meet that challenge.

CHAPTER 2
THE NAVAL ACADEMY AND
THE FAITH OF MY FATHER

From the moment I stepped onto the grounds at the United States Naval Academy in Annapolis, Maryland, on July 1, 1957, two things became abundantly clear: The first was that there was no doubt just who the boss was, and it was not me. The second was that the Navy was about to take away all my God-given rights and give them back as privileges, one at a time, grudgingly and sparingly, over the next four years.

I really had no sense about what was in store for me when I got to the Academy. I did not talk to anyone who had been a Midshipman who might have given me a sense of what it was going to be like. Even if I had, I do not think anyone would have been able to adequately describe what was in store for me over the next four years—especially during those first two months known as Plebe Summer.

The term Plebe Summer is deceiving. The only thing "summery" about July and August at the Academy is the heat and humidity that wrapped the place in a cocoon of hot, moist air oozing off the Severn River. Summer implies time off, vacations, and good

27

times; Plebe Summer is 60-plus days of hell in a very small place, with nowhere to run, nowhere to hide, and nothing to do but take what the upperclassmen dish out—because, in the end, all Plebes volunteered for this.

I just remember thinking on more than one occasion that summer, "What have I gotten myself into and how can I get myself out of this?"

Induction Day, or I-Day, is pretty much a blur after all these years. I remember checking in and at some point, taking the oath in which I pledged to support and defend the Constitution of the United States against all enemies, foreign and domestic. Then came hours of nervous Plebes standing in any number of lines where uniforms and gear were handed out and stuffed into over-sized white laundry bags while we were yelled at by upperclassmen because we were not doing it properly or fast enough to suit them. At the end of the day, I was just one of 1,148 Plebes dressed in Navy white jumpers and trousers with the iconic, blue-trimmed Dixie cup sitting on our freshly shaved heads.

What became clear quite early on was that Plebe Summer was designed not only to begin transforming us from sloppy civilians into disciplined Naval officers; it was also the start of a process to weed out those who were not able to meet the Navy's physical, psychological, and academic demands.

If a Plebe were out of shape physically, the upperclassmen responsible for his care and feeding were sure to give him a little extra attention to get him properly motivated—or else. If he was not up to the demands psychologically, the upperclassmen exploited it. And, if he was a little hardheaded or a bit of a wise guy—as I had a tendency to be at times—they tried to knock that out of him pretty quickly. I tried as best I could to keep my mouth shut and my head down those first few weeks.

The physical demands of Plebe Summer were not particularly difficult for me. My high school football background and work on the farm had gotten me into decent physical shape before I got to Annapolis. The military aspect—learning to march and salute—also came rather easily because of the time I spent in the Michigan Army National Guard. Psychologically, though, Plebe Summer was a challenge. I was still a rather impressionable, if not a bit naïve, 18-year-old kid from a small town in Michigan and had never been through anything like this before.

According to the upperclassmen, nothing the Plebes did was ever done fast enough. Shoes never were shined well enough. The uniform was never immaculate enough. Answers to questions never came quickly enough or with the right amount of conviction. It seemed neither I nor any of my fellow Plebes could do much of anything the way the Navy wanted it done that summer.

The upperclassmen, also known as the *cadre* or *detailers*, were on our backs from reveille at O-Dark-Thirty to lights out. They tried to get into our heads, and some of my fellow Plebes found themselves allowing those upperclassmen to live in their heads rent-free most of that first year. But, as I later came to realize, the intent of the *cadre* was not merely to harass us for the sake of harassment—although it seemed that way at times—they were also testing our limits and trying to strip us of all vestiges of civilian life so we could be re-made into naval officers. They were teaching us that as individuals we were of little value to the Navy; instead, we were learning to function as a team the Navy way.

The upperclassmen were irritating, aggravating, and so full of themselves that it was always fun to see one of them get his comeuppance every so often. One day, I was at chow with a classmate named Dick Kibbe, when an upperclass-

man at our table said, "Mr. Kibbe, what does your father do?"
"My father is in the Navy, sir," Kibbe replied.

"What rank is he?"

"He's a captain, sir." In the Navy, a captain is the equivalent of
a full colonel in the other three services, which means he has been
around for a while.

"Has he ever had command of a ship?" the upperclassman went
on.

"Yes, sir."

"I bet it was a garbage scow."

"I guess you could say that, sir," Kibbe responded. "It was the
USS Forrestal, sir."

The upperclassman's mouth dropped open. At that time, the
Forrestal was the Navy's newest aircraft carrier. Not only that, but
the senior Richard Kibbe was a naval aviator awarded the Navy
Cross—second only to the Medal of Honor among Navy battlefield honors—while fighting around the Philippines during
World War II. The chastened upperclassman turned around and
slunk off.

Kibbe was not the only son of a high-ranking naval officer at
the Academy in the Class of 1961. Those Plebes did not necessarily get any better or any worse treatment because they were the
sons of senior officers; for that to happen, they would have had to
have been the son of the Chief of Naval Operations. Nor did they
do any better than the rest of us either academically or militarily.
From my experience, I think they did worse than most because
some of them were not there by their own choosing: They were
there because their fathers wanted them to be there, and they were
expected to follow their fathers into the Navy.

During Plebe Summer, there was so much to absorb, so much
to remember, and so many new things to learn that at times it

seemed impossible to keep up. *Reef Points*, a 200-plus page handbook given to every Plebe, only contributed to that sense of being force-fed more information than any human could learn in two months and be expected to retain.

Reef Points is everything Navy. It has information about chain of command, Navy history, Navy customs, Navy traditions, Navy ways of signaling and telling time, and Navy language of the kind that can be repeated in public. (The other kind we learned in the residence hall, the mess hall, or the wardroom if we didn't already know it.)

In Navy language a floor is the deck, a wall is the bulkhead, the ceiling is the overhead, a hat is a cover, a hallway is a passageway, a door is a hatch, and a bathroom is the head.

We had to learn to tell time by the Navy system of using bells to signal every half hour in each of six watches over a 24-hour period. We also had to learn how to recognize the signal flags Navy ships use. The only one I remember after all these years is "Bravo Zulu," meaning "well done."

We also had to memorize Morse code. I was reasonably good at Morse code but never once used it during my time in the Navy. I guess learning it was a nod to the "old" Navy. I doubt if many sailors these days know what Morse code is.

Reef Points also had a section called "Table Salt" that contained what to the normal person would be nonsensical sayings we had to memorize and then regurgitate whenever an upperclassman asked us a certain question. I am not sure what these sayings did for us, but we were expected to know them by heart.

If an upperclassman stopped you and asked, "How long have you been in the Navy?" the appropriate response was, "All me bloomin' life, sir! Me mother was a mermaid, me father was King Neptune. I was born on the crest of wave and rocked in the cradle of the deep. Seaweed and barnacles are me clothes. Every tooth in me head is a marlinspike; the hair on me head is hemp. Every bone in me body is a spar, and when I spits, I spits tar. I'se hard. I is, I am, I are!"

Or, "What time is it?"

"Sir, I am greatly embarrassed and deeply humiliated that due to the unforeseen circumstances beyond my control, the inner working and hidden mechanisms of my chronometer are in such accord with the great sidereal movement with which time is generally reckoned that I cannot with any degree of accuracy state the correct time, sir. But without fear of being too greatly in error, I will state that it is about __ minutes, __ seconds, and __ticks past __ bells."

If an upperclassman stopped by the table at mealtime and asked, "How's the cow?" he was referring to the milk served in large metal pitchers. We would have to dutifully respond, "Sir, she walks, she talks, she's full of chalk. The lacteal fluid extracted from the female of the bovine species is highly prolific to the (approximate number of milk glasses remaining in the pitcher) nth degree!"

And, if an upperclassman yelled, "Fire in the paint locker!" in the mess hall, the Plebe closest to the milk pitcher had to throw it under the table. That never made any sense to me because not only was it a waste of good milk, it ruined everybody's shoes and we would have to spend hours cleaning and polishing them.

Things only got worse for us Plebes when the full Brigade returned in September. Then, instead of having a few hundred upperclassmen telling us what to do, how to do it, and how badly we were doing whatever it was they wanted us to do, now there were 3,000 of them doing it. It was the difference between being chased by one dog or a pack of dogs.

Everyone we met was addressed either as "sir" or "ma'am," depending on their gender. If you did not address someone with "sir" and he was an upperclassman, he would likely pull you up short and tell you, "Your ass is grass and I have the key to the lawnmower shed!"

We also saluted just about anything in a uniform; I don't know how many times I saluted the Coca-Cola delivery man. We even saluted the civilian gate guards, the guys we called "Jimmy Legs," a name derived from old Navy slang for the Shore Patrol.

Home for the Brigade of Midshipmen was then and still is today, Bancroft Hall, sometimes referred to as "Mother B." It housed our rooms, the mess hall, dry cleaners, bowling alley, pistol range, laundromat, uniform store, cobbler shop, and post office, among other useful offices and establishments. It has undergone some renovations since I was there and for some time now has had air conditioning, a luxury unknown to us in the late 1950s. Lack of air conditioning made for some nights almost as hot and sweaty as the days during Plebe Summer.

The Brigade consisted of two regiments, each with three battalions with four companies in each battalion. Each company was like a standard military company of about 130 Midshipmen. I was assigned to the 1st Regiment, 1st Battalion, 4th Company. Each battalion had a commanding officer who was either an Army or Marine Corps lieutenant colonel or a Navy commander, all of which are the O-5 rank. A Navy lieutenant commander or Army or Marine Corps captain or major commanded each company.

Once the Brigade returned to the Academy in September and classes began, the feeling that I was overwhelmed and under-prepared for this type of life only increased. The Plebes were constantly harassed by the upperclassmen and our repeated failings in all aspects of our military bearing were constantly pointed out to us and to any other upperclassmen within hearing range. Academically, the pressure also increased. The year I spent at Northwestern Prep prior to entering the Academy emphasized math and science, but at the Academy the difficulty of those subjects went to an entirely new level.

That Plebe year everyone took the same classes, which included courses such as English, seamanship, ordnance, shipbuilding, government, chemistry, Naval history, and leadership. The only elective was a foreign language. Although I grew up in a family where Polish was spoken regularly, I did not speak the language, and it was not among the course offerings. I could have taken Spanish or French or German, but a number of the Plebes had taken those languages in high school, and I figured if I got into class with them, I'd be 'way behind from the start. I also knew Chinese was not a language I wanted to tackle.

So, I chose Portuguese. Why Portuguese? Since almost nobody took Portuguese in high school, that seemed the best way to go because everybody else would be starting at the bottom, like me. I took Portuguese my first two years at the Academy but never really learned to speak it well enough to say much beyond "Hello," "Goodbye," "Where's the bathroom?" and "Bring me another beer."

The two subjects with which I struggled greatly throughout my four years at the Academy were mechanical drawing and literature. I'm not good at visualizing something and trying to reproduce it, and that's what mechanical drawing is all about. My mechanical drawing class was taught by a Navy lieutenant who was an Academy grad. One day he took me aside and said, "Mr. West, I predict that one day you are going to come back here as an instructor."

"Why is that, sir?" I asked.

"When I was here, I almost flunked this class and you're just like me, almost flunking."

My struggles with mechanical drawing meant I had to spend a

lot of time in what is referred to as "Stupid Study." My grade-point average was below expectations—a 2.7 or 2.8 if I recall correctly—so I had to take an extracurricular class two times a week to get up to speed.

I've also never been a fan of the Humanities. In my senior year we were studying what we were told were great works of literature. One of them was Dante's *Inferno*. One day the instructor said to me, "Mr. West, what did Dante say to you when he wrote..." and he quoted something from the poem. I rattled off some nonsense and the instructor looked dumbstruck.

"That's not what Dante told you," he said.

"Yes, it is, sir," I responded.

"No, it's not," he countered.

"Sir," I replied, "you asked me what Dante told *me*, not what he told *you*."

I didn't get into trouble with the instructor for being a wise guy, but he was not particularly happy with my answer. I did not really care, though, because I never liked that class.

Throughout my four years at the Academy, the academic demands remained much the same. The only thing that changed was the amount of time I had to study. As a Plebe, I had virtually no time to study because there were too many other demands on me. Either I knew it, or I didn't. By the second year, also known as the Third-Class year or "Youngster Year," I had more time to study because I didn't have all those upperclassmen harassing me. In the final two years, much of the focus was on succeeding academically.

In addition to the academics and the military aspects of the Academy, we were required to participate in one sport every spring and fall; it could be an intercollegiate sport or an intramural sport. My football skills were not such that I was recruited by the Academy for its team, nor did I consider trying out as a walk-on because at that time the Navy football team was still considered one of the nation's elite collegiate programs. In fact, one of my classmates was Joe Bellino, who won the Heisman Trophy in 1960. He later went on to play in the old American Football League for the Boston Patriots, and in 1977 was elected to the College Football Hall of Fame.

Joe was an extremely likeable guy and did not have the sort of ego you might expect of someone who got the amount of publicity he did. Academically, though, Joe struggled. His football ability made up for what he lacked in the classroom. We used to say that if you got a grade better than Joe in any class you were going to stay at the Academy. I never would have been able to handle both the academic demands plus the practice demands of football.

My sport of choice was something called fieldball. It is a strange combination of soccer, team handball, football, and lacrosse where the object is to kick, throw, or head a soccer ball into a lacrosse net. Unlike lacrosse, there are no sticks with which to whack people, and the players do not wear helmets or pads.

Fieldball teams have nine players each (eight field players and one goalie). Players are required to keep two hands on the ball while running with it but must bounce it once to themselves at any point during their possession while the opposing team tries to gang-tackle the guy with the ball. The game is composed of two 30-minute halves, and goals are worth two points each, except for penalty shots and shootout goals, which are worth three and two points, respectively. The game is rough, dirty, and can be bloody. Maybe that's why the only two places fieldball is still played are the U.S. Naval Academy and the New York State Prison. I'm not sure whether that says something about the Academy or the prison, but the game enabled me to get rid of a lot of aggression and frustration. My senior year I captained the fieldball team that won the Brigade championship, a sterling achievement among those familiar with the game.

Plebe Summer and that first year were so demanding on me academically and psychologically that had I been able to, I would have walked out the gate and never looked back. Plebe Year is like a year-long boot camp with academics. At that time, though, you had to be 21 to leave the Academy of your own accord. If you were under 21, you had to get permission from your parents to leave.

Two things kept me going, though.

The first was that every time I started feeling sorry for myself and thought about trying to convince my father to give me permission to quit, I recalled a conversation I had with a sergeant in the Michigan Army National Guard after he learned I was going to the Naval Academy.

"Dave," he said smugly, "you will never make it at the Naval Academy."

I told myself I would not give him the satisfaction of giving up on myself.

Still, I don't know how many times I wrote to my father virtually begging him to sign the papers so I could leave. "This isn't going to work," I wrote. "You've got to get me out of this place."

The second thing that kept me going was my father. He invariably would write back and say something like, "You'll be home for Christmas soon, and you'll have two weeks to spend with your friends. We'll talk about it then." Or, "Well, remember you're going on a cruise in a couple of months. Get through that and we'll see what things are like at the end of the summer."

My father's letters were always full of support and encouragement. Had it not been for him, I'm not sure I would have made it. He provided the moral support I needed to get through that Plebe year. I started to think that if he thought I should stay here, maybe

I ought to stay. He kept telling me good things were on the horizon and maybe I should stay to see what they were.

I think some of the motivation he gave me was his way of saying he wanted me to succeed at something he never had the opportunity to do. I am sure he was proud of the fact that his son was at the U.S. Naval Academy and did not want me to throw away this opportunity. He had more faith in me than I had in myself at the time, and I'm thankful I didn't disappoint him.

Except for Plebe Summer, the worst part of that first year came after I returned to the Academy from Christmas vacation. That post-Christmas period was referred to as "The Dark Ages." It was miserable there after Christmas: You never saw the sun and half the time you saw snow. The other half of the time you saw the pigeons crapping on you, and there were a lot of pigeons around the Academy. Had it not been for my father's upbeat and encouraging letters, those days would have been unbearable. I pressed on and eventually May 1958 rolled around, which meant the end of Plebe Year and the transformation into our Third-Class year and our first cruise.

First, though, was the traditional capping of the Herndon Monument that unofficially marks the end of Plebe Year. Anyone familiar with the Academy and its traditions is aware of the Herndon Monument, a 21-foot-high gray granite obelisk near the center of campus. The monument is in memory of Navy Captain William Herndon who, in 1857, decided to go down with his ship rather than try to save himself.

Plebes are no longer considered Plebes once they gather as a class and one member scales the monument to replace a Dixie cup cover with an upperclassmen's cover. The only impediment to getting this done quickly and easily is that the monument is slathered in grease, oil, lard, butter, and anything else that makes getting to the top a rather difficult undertaking.

The idea is to get the Plebes to work together with larger and stronger men at the bottom supporting lighter, more agile classmates in something of a human pyramid so that one Plebe is able to reach high enough to knock the Dixie cup off the top and replace it with an upperclassman's cover. I was not particularly big, but I was one of the Plebes at the bottom of the scrum providing the foundation so someone could get up there and get that damned Dixie cup off the top of the monument. Tradition says that the Plebe who reaches the top will become an admiral, but I'm not sure that has ever happened.

The Academy did not start keeping records of the time it took each class to complete the Herndon Monument capping until the following year. The class of 1962 did it in 12 minutes. The fastest time ever recorded was 90 seconds by the Class of 1972, but that was during a period when Academy officials prohibited the use of anything greasy or slippery being placed on the monument, a distinct advantage for that class.

After the Herndon capping, the three classes remaining at the Academy are sent to various Navy ships and installations to get a working introduction to what life in the real Navy is like. Some go on cruises, some are sent to naval air stations (NAS), and some are sent to train with the Marine Corps.

The summer following the end of my Plebe Year in June 1958, I had the good fortune to be selected for a cruise to Europe and several countries in the North Atlantic Treaty Organization (NATO). Even better, my assignment was on the aircraft carrier *USS Intrepid,* giving me my first opportunity to get a close look at carrier flight operations. Watching the intricately timed ballet of flight operations on the pitching deck of the carrier only further stirred my desire to become a pilot and to be one of those aviators who had the nerves and the skill to do this special kind of flying.

Just watching it was exciting. Doing it, I was sure, would be exhilarating.

By no stretch of the imagination was this a pleasure cruise, however. Nor were the Midshipmen considered Navy officers. We were deck hands, learning about the Navy at sea from the point of view of an enlisted sailor. While working on the ships we usually wore the traditional enlisted sailor's working uniform of blue bell-bottom dungarees, blue denim shirts, white Dixie cups, and black boots. We did a lot of paint chipping, repainting, cleaning up, and just about every menial job that can be performed on a ship. I suppose much of what we did was designed to make us glad we were going to be officers and not enlisted sailors.

During that cruise, I spent a good deal of time working in the boiler room of the *Intrepid*, where I learned that I certainly did not want to end up as an enlisted sailor. One of the regular crewmembers in the boiler room at that time was a three-striper, a Seaman. He was in his mid-to late-20s and had been in the Navy eight or nine years but was still only an E-3. One day I asked him, "Are you going to take the upcoming test for (Petty Officer) Third Class?" It was the next highest rank and meant a bump in pay.

"Nope," he replied.

"Why not?"

"Well," he explained, "if I make Third Class, I'll go to the bottom of the watch list and I'll get all the bad watches. As a senior three-striper, I get my pick of watches."

I thought to myself, "Well, here's a guy with a lot of ambition!" I was surprised he had not been released from active duty, which is what the Navy would do nowadays with someone who spends too much time at one grade, especially if he had no desire to advance.

Although there was a lot of work to do on the ship while we were at sea, once we hit port, we were able to kick up our heels a bit, and act like, well, sailors on liberty. It wasn't always pretty, but it usually was a lot of fun. I am still amazed at some of the things we did while managing to avoid getting picked up by the Shore Patrol.

I had developed a friendship with Denny Moore, who was from the Minneapolis area and had attended Northwestern Prep with me the year before. Like me, Denny wanted to fly jets, so we had that bond in addition to wanting to have a good time whenever we could.

When we stopped in Oslo, Norway, in July, he and I went to a little town north of Oslo, where a number of American students took classes during the summer. We went to a little bar near the school, I think to check out the girls. At one point during the evening, I told him I had to go to the bathroom.

"Well, leave some money because it's your turn to buy," Denny said. I gave him a $20 bill and said I would be right back. When I returned, the three-foot wide table was completely covered in wine bottles.

"What the heck happened, Den?"

"I ordered some wine, but the waitress said she didn't have any change."

I did not realize it, but they were charging only 50 or 60 cents for a bottle of wine and the waitress had given us $20 worth of wine. I think we ended up sharing most of it and leaving the remaining bottles on the table.

Ten days later, we stopped at the port of Rotterdam in The Netherlands. Instead of staying there and wreaking havoc on the local economy, Denny and I decided to a take the train to Brussels, Belgium, where the 1958 World's Fair was in progress. Neither Denny nor I had ever been to a World's Fair, and this was the first one since the end of World War II.

Although the pavilions built by various countries for what was known as "Expo 58" were quite impressive, one of the more interesting things we saw was the Heineken beer stand where we could get a free beer. Denny and I would go in and have a beer, then leave for a while, then come back, and have another beer. About the third time we went back, the guy serving the beer looked at us and said, "You guys look familiar. Have you ever been here before?"

And we're like, "No, this is our first time here." The guy eyed us suspiciously but gave us more beer.

How Denny and I did not get into big trouble was due, I think, to just sheer, dumb, luck. During a stop in Barcelona, Spain, on another cruise one year, we were in a cab heading back to the ship when Denny accidentally dropped a lighted cigarette in the crack of the back seat and could not get it out. Before we knew it, the cab started filling with smoke. The driver slammed on the brakes, and we hopped out, walking off before he could figure out what had happened!

Another time we were in a bar drinking Spanish whiskey, which was really bad whiskey. Denny looked at me and said, "Why don't we buy a bottle of decent American whiskey?" I agreed and we asked the waiter to bring us a bottle. No sooner had the bottle hit the table than a bunch of Spanish hookers looking for a decent drink surrounded us. Denny looked at them, pulled the bottle close to him, and refused to give any of them a single sip.

At the end of that Third-Class cruise, I was scheduled for one final surgery on my right hand at Bethesda Naval Hospital, now known as the Walter Reed National Military Medical Center. That last surgery was mostly cosmetic; doctors took out some of the skin that had built up on my palm and cut down on the webs between my fingers, making the hand look like any other right hand. Most people at the Academy did not even know my hand had been an issue at one time.

During the two weeks I was at Bethesda, I met another Midshipman who also was recuperating from an injury, but his was a result of football. His name would not mean much to today's sports fans, but in the late 1950s Bob Reifsnyder was one of the top players in the country. At 6-foot-2 and 250 pounds, he would be considered undersized for major college football these days, but in 1957 he was a hulk of a man. He played offense and defense as a tackle and at the end of the 1957 season—in which Navy went 8-1-1 and beat Rice in the Cotton Bowl—he was named to the All-American team and awarded the Maxwell Trophy. (The Maxwell is somewhat similar to the Heisman in that it is presented to the college football player considered the best in the nation that year. The only difference between the two is who votes for them. Heisman voters are members of the media and past Heisman winners; Maxwell voters are members of the media and college football coaches. For Reifsnyder to be awarded the Maxwell Trophy was very impressive because he was the first lineman ever to receive it).

Bob was at Bethesda for surgery to repair his Achilles tendon, which he tore just prior to the 1958 season. I no longer recall how we ever got to talking, but I soon discovered that Bob was something of a free spirit, much like me, and was not averse to challenging the Academy's strict regulations.

While we were recuperating, we occasionally sneaked out of Bethesda and hopped a cab for the 15-mile ride to College Park, Maryland, where we trolled the bars around the University of Maryland campus looking for girls. When we got back to the Academy following our recuperations at Bethesda, Bob and I continued our unauthorized forays, sometimes ending up on the waterfront in Baltimore where the tough longshoremen hung out.

Nobody messed with Bob because of his size and, since I was with him, nobody messed with me.

We called those little escapes from the Academy "Going Over the Wall." Had we been caught, we would have been severely disciplined and possibly expelled. That probably would not have been a big deal for Bob because his Navy career was going to end after graduation as a result of that tendon injury. For me, it would have been more of an embarrassment to my father than to me, but I liked the challenge of getting away with something. Since Bob was two years ahead of me, he knew all the tricks about getting in and out of Bancroft Hall and the Academy unnoticed by the upperclassmen or the Jimmy Legs.

One thing about Bethesda Naval Hospital that caught my attention—and I thought rather odd—was a small pond that was brimming with golf balls located in front of the main entrance. The hospital was just a few hundred yards from one of the holes of the Columbia Country Club golf course and, apparently, more than a few not-so-accurate golfers hit shots that ended up in that pond, and nobody had bothered to retrieve their golf balls.

Thinking I could make a lot of money on golf balls if I were able to get my hands on them, I recruited the brother-in-law of one of my friends at the Academy to go over the wall with me one weekend night. When we got to the pond, we took off our pants and our shoes and socks and waded into the water in our skivvies, picking up golf balls.

We must have pulled 100 to 150 golf balls out of the pond. As I was driving my friend's car back to Annapolis, I was so giddy about what we had done I didn't notice how fast I was going. All of a sudden, I saw the lights of a police cruiser in the rear-view mirror and pulled over.

It was the Maryland State Police and as the officer shined the light in our car, he must have been more than a little surprised by what he saw: two guys in their underwear with dozens of golf balls in the back seat. He looked at us and irritably asked, "What the heck's going on here?"

"We're from the Naval Academy and were just over at Bethesda," I said, thinking that explanation might buy us a get-out-of-jail-free card.

"You're telling me you're from the Naval Academy?" he asked, even more skeptical.

"Yes, sir," I replied.

"What were you doing over at Bethesda?"

"We were getting golf balls out of the pond," I said, reaching into the back and pulling out a few for him to see. "Here, do you want a few?"

He looked at us again and then growled, "Get out of here!"

Although I never got into any trouble as a result of my over-the-wall extracurricular activities, I was not so lucky inside the walls. The upperclassmen always seemed to find me when I did or said something they felt was inappropriate.

One upperclassman I seemed to keep getting crosswise with was Jack Funderburk. He was a second classman (junior) my Plebe year and was in the same company, so we ran into one another quite a bit—and not always in circumstances that ended up well for me. He also was a few years older than I was, having spent a year at The Citadel and three years at the University of Maryland before coming to Annapolis.

One day during that Plebe year, a group of us were marching to class, and the unit in front of us was goofing off and slowing us down. Getting to class on time was a big deal because if you had a test and got there late, you never got any extra time to finish the test, so it behooved you to get to class on time.

"Get those so and so's out of there!" I shouted from my spot in the formation to no one in particular. Breaking ranks like that is considered a rather serious breach of military discipline, and the Academy and the upperclassmen frown on such conduct.

As soon as the words came out of my mouth, Jack Funderburk turned around to look for the culprit. He asked one Plebe if he had made the remark and he hastily replied, "No, sir." Then he turned to me and said, "West, did you say that?" Lying is an even more serious breach of the Honor Code than being disrespectful in ranks so I responded, "Yes, sir."

Funderburk came up close to me, looked me in the eye, and said sternly, "Come around."

Now, to the average civilian, that might have sounded as if Funderburk was telling me to shape up and get my act together. At the Academy, it means something completely different, and it is one of those things you never want to hear, especially as a Plebe.

At the Academy, the command "come around" meant that I had to be in front of Funderburk's room at least 15 minutes before each meal. I was required to stand there at attention and shout out, "Chow time! Sir, there are 15 minutes to chow time and the menu is..." and I would have to recite every item on the menu for that particular meal, whether it was breakfast, lunch, or dinner. Every 60 seconds I would have to do it again. "Chow time! Sir, there are 14 minutes to chow and the menu is..." and so on until it was time to go to the mess hall.

Funderburk also popped out of his room from time to time during those recitations and asked questions that he knew I had no answer for and so I would tell him, "I'll find out, sir." You would never say, "I don't know." That implied you did not care enough to look it up.

Funderburk had one roommate who was an opera fan, so this

guy also started asking me questions about opera. I knew more about brain surgery than I knew about opera, so I was completely lost. I eventually enlisted classmates to go to the library to look up the answers so I could memorize them for the next time I had to stand at Funderburk's door. That bit of harassment lasted from before Christmas my Plebe year until almost June Week.

Funderburk was not all harassment, though. One Saturday night late in my Plebe year, George Mays, one of Funderburk's roommates, invited me and another Plebe to play bridge in their room. I was seated with my back to the door. As I previously mentioned, I am an intense card player—whatever game I'm playing and whomever it is I'm playing with, I don't mess around when I play cards.

All of a sudden, the door opened, and a voice called out, "How are you doing, Mr. West?"

Focused on the cards, I just casually replied, "Okay."

"Goddamn you! Don't you ever learn?" the voice thundered.

I turned around in my chair. It was Jack Funderburk. My first thought was, "Oh, shit!"

Funderburk looked at me and said, "Mr. West, I flunked one of my classes. I'm slated for a re-take. If I pass, I'm here for another year. If I don't, I'm gone. From now on, though, you can call me Jack, because anyone with balls as big as yours has earned that."

And then he spooned me, which means he shook my hand. He held out his hand and said, "My name is Jack." That meant I did not have to call him "Sir," any longer. I could just call him "Jack."

People have asked me when I felt I had turned the corner at the Academy, and whether once I got into my second year if things settled down for me. Truthfully, I never felt I had turned the corner until June 7, 1961, the day I graduated. I always felt as if I was on the verge of not making it either because I might fail academically or that my wise-guy remarks to the wrong person might just earn me a one-way ticket back to Ironwood and a life sentence in the iron ore mines. More than once during my Plebe year, I did something that got me "fried," which is Naval Academy-speak for getting demerits and being forced to run around the field house at 5 a.m.

I also left my mark on the Academy—literally. The upperclassmen liked to make the Plebes stand at a stiff brace for a half hour or more in front of their metal lockers if we did something of which they did not approve. I do not recall how much time I spent doing that, but it was enough so that when I went back to the Academy for a football game in 1964, three years after my graduation, I could still see the outline of my sweat on my old locker.

I was overjoyed to have that Plebe year behind me; my yearlong boot camp with academics was finally in my rear-view mirror.

* * * * *

Youngster Year, or the Third-Class year at the Academy, provided a small measure of relief after the pressures of the Plebe year. There never were a lot of privileges at any time at the Academy, but that second year we could at least go into Annapolis and go to the movies on Saturdays. That also was the year I went over the wall with Bob Reifsnyder on several occasions. But, beginning that year, much of the harassment ended; all I had to do was stay out of trouble, keep my mouth shut, and study.

There were occasional incidents that helped relieve the tedium of study/class/drill. One in particular I remember occurred in the mess hall. I am not sure what year it was, but it was in the pre-Civil Rights era.

During my years at the academy, the majority of the mess stewards were either black or Filipino and the majority of Midshipmen were white. One day the entire Brigade of 3,800 Midshipmen marched into the mess hall for lunch only to discover lunch was not ready.

The head of the mess hall was a Navy commander. He went to the chief steward and said, "What's going on here?"

"We're not going to serve those white boys anymore," he was told.

The commander looked at the other stewards and asked if they all agreed.

"Yes, sir," one responded, "we ain't gonna serve them no more."

The commander had the sergeant-at-arms with him, turned to him and said, "Sergeant, go over to the armory and get a .50-caliber machine gun and some ammo."

The chief steward asked him, "What are you gonna do with that?"

"Well, you know this is mutiny, so I'm gonna have to shoot you all."

You've never seen people move so fast in your life to get lunch ready.

It may not have been politically correct in today's world, but back then that was just the way it was.

A certain routine developed outside the normal academic year. Every late November or early December, we boarded buses and headed to Philadelphia for the Army-Navy game at what was then known as Municipal Stadium on South Broad Street. Then we would stand in the cold or the rain or the snow for a few hours and cheer on the Midshipmen. After the game, we usually had a few hours to spend in downtown Philadelphia before catching the buses back to Annapolis.

We were given two weeks off at Christmas to spend time at

home with the family, and each summer we had roughly a month off after the cruise. Several of those summers, I went home and worked for the county on a road repair crew. My ex-girlfriend's father, Gene Velin, the mayor of Ironwood, helped me get that job, even though I was no longer dating his daughter.

At the end of my second year, in the summer of 1959, I was sent to the Naval Air Station in Jacksonville, Florida, for what is referred to as the aviation cruise. Other Midshipmen went to other naval aviation facilities as part of their continuing education into all things Navy. While at Jacksonville, I was able to talk to pilots and fly in some of the Navy's aircraft, particularly the T-34B, which was the Navy's primary trainer at that time.

My class was the first that did not have the opportunity to actually fly an airplane during our time at the Academy. Until 1957, Second-Class Midshipmen (juniors) were able to fly with an instructor in the Naval Aircraft Factory N3N, a two-seat, open-cockpit biplane lovingly called the "Yellow Peril" because of its bright yellow color. It was replaced by the T-34B, made by Beechcraft, but those of us who became pilots would not get to fly that particular aircraft until we actually got to flight school.

The more I was around airplanes and aviators, the more I knew I wanted to become a pilot: the only question was whether to become a Navy pilot or a Marine Corps pilot. For a brief time, I toyed with the idea of flying for the Marine Corps. At one point, a Marine Corps general who was a pilot gave a presentation at the Academy and told us, "A lot of people resent us because we get flight pay. I'm here to tell you pilots don't get any more money than anyone else; they just get it faster."

But, at that time most Marine Corps squadrons were land-based. When I realized that fact, I said to myself, "I'm not going to do that." I wanted to fly off carriers.

In those days, if you were qualified and you wanted to go into aviation coming out of the Academy, you got aviation because the Navy was in need of aviators. I think the brass saw Vietnam coming down the road and believed there would be a need for naval aviators. The French had gotten their butts kicked in Vietnam in the early 1950s, and the country had been divided at the 17th Parallel into North and South, with the United States aligning itself with South Vietnam politically and economically. The United States had advisors there, and I think the people in power in Washington realized it was only a matter of time before we would become more involved militarily.

One of our assignments in a navigation class was planning an invasion of Vietnam. The instructors called it "Korea" at the time, even though that war had come to an unsatisfactory stalemate some years earlier. But, I guess they did not want it to leak out that strategic planners in the military were starting to look at Vietnam as our next conflict.

That third year, known as the Second-Class year, was noteworthy only for that it was not particularly noteworthy, unlike my other years at the Academy. The routine of study, drill, sports—but mostly study—carried me through that year, as did the continued encouragement of my father in the regular letters he wrote to me. Whenever I felt as though the Academy was getting the best of me and I let my father know, he would send a letter that helped me focus on the future, not on whatever I was feeling at that particular moment.

By the time we finished our third year at the Academy and transitioned from Second Classmen to First Classmen, the only thing everyone wanted to do was get the year over with and get on to what we would be doing for the rest of our Navy careers. We had a series of speakers come in to talk to us about various aspects of the Navy, and one of the more memorable was a presentation by Vice Admiral William F. Raborn, Jr., a submariner who was instrumental in the development of the Polaris missile. His presentation was memorable not for what he said, but for what he did *not* say. After his presentation, he opened the floor to questions. The first question was, "Admiral, how deep can a nuclear submarine go?"

The admiral thought for a moment and then responded, "Pretty deep."

The next question was, "How fast can it go?"

"Pretty fast."

It was then I knew we weren't going to get much information from the admiral because most of what he knew about submarines was classified. I said to myself, "We'd better stop asking questions because he's not going to say anything substantial."

Another focus looming on the horizon for the United States was space exploration, so it wasn't a coincidence that my class chose a rocket ship as our class crest rather than the traditional ship motif.

My final year at the Academy seemed to drag on longer than my Plebe year. All I wanted was to be finished with it and get to my first duty assignment. I was so eager to graduate and get away from Annapolis that I barely remember June Week, when Duke Ellington and Julie London came to perform. At the time, both were considered among the nation's most popular entertainers.

Graduation day finally arrived on June 7, 1961. My parents came in from Rhode Island and were joined by my uncle, Carl Heczko, his wife, Leah, and their daughter, Shirley. Uncle Carl worked for the forest service and was on temporary assignment in Washington, D.C., so they lived close enough to drive up for the ceremony.

I was especially happy my father was able to make it to the ceremony. Without his encouragement and support in the numerous letters he had sent me over the previous four years, I don't think I would have made it through. He refused to give up on me when I was thinking about giving up on myself. He was not the reason I had come to the Academy, but he was a major part of the reason that I made it through those four years.

Graduation ceremonies at the Academy often are held outside to accommodate the large number of family and friends of the graduates who wish to attend and was moved inside usually only in inclement weather. Our 1961 graduation, though, was held in the Field House, in part, I think, because it was a steamy, 80-degree day and, in part because our commencement speaker that year was President John F. Kennedy, Jr.

Seated in front of Kennedy that June morning, resplendent in our white dress uniforms, the class of 1961 numbered just 788—360 fewer than had come to Annapolis for Plebe Summer in 1957.

Although Kennedy was not a Naval Academy graduate, he had served as a naval officer during World War II and earned distinction for his role as commander of PT-109. His boat had been cut in two by a Japanese destroyer in the Pacific, and Kennedy was instrumental in saving 10 of his crew and keeping them safe until they were rescued. A back injury Kennedy suffered in the incident bothered him for the rest of his life.

In his speech, Kennedy was relaxed and somewhat self-deprecating, telling us, "In the past I have had some slight contact with this service, though I never did reach the state of professional and physical perfection where I could hope that anyone would ever mistake me for an Annapolis graduate.

"I know," he continued, "that you are constantly warned during your days here not to mix, in your naval career, in politics. I should point out, however, on the other side, that my rather rapid rise from a reserve lieutenant, of uncertain standing, to Command-er-in-Chief, has been because I did not follow that very good advice."

Kennedy thanked us for our commitment to the country at a time of great uncertainty and for the sacrifices we were about to make to help defend the country. What he did not know then was just how much circumstances would change over the next few years — not only for him personally, but for the government, the military, and the people of the United States.

When the speeches were over and we were finally ensigns in the Navy or second lieutenants in the Marine Corps, we tossed our covers into the air and rushed to celebrate with our families.

Not only was I relieved to have graduated from the Academy, I was also pleased because my first choice of assignments, aviation, had been approved. I was on my way to Pensacola, Florida, where the Navy would make the final decision as to which type aircraft I would learn to fly.

Would it be a prop-driven transport aircraft?

I certainly did not fancy the idea of being a delivery man for the Navy.

Would it be a helicopter?

I had no desire to fly helicopters. I've always considered heli-copters to be an accident looking for a place to happen.

Jets were foremost in my mind, but I would have to wait a few more weeks to find out if the dream I had been nurturing since I was a youngster would actually come true.

CHAPTER 3
SHOES OF BROWN, WINGS OF GOLD

Unless someone has served in the United States Navy, it may come as a bit of a surprise to learn there are two distinct branches of the sea service. There is the "Black Shoe Navy" and the "Brown Shoe Navy." There actually is a third branch of the Navy family—the U.S. Marine Corps; however, we more civilized Navy types consider them unruly cousins whom we try to avoid unless we get into a particularly nasty fight. At that point, the Marines' constant gung-ho-ism can come in handy.

Those who serve in the Black Shoe Navy make up the overwhelming majority of the service. As the name implies, the normal footwear for that side of the family consists of black shoes and, occasionally, black boots. The Brown Shoe Navy is a rather exclusive club, thus the choice of separate footwear. Those brown shoes—and brown flight boots—are reserved for naval aviators.

As I headed to Naval Air Station Pensacola in Florida's Panhandle in July of 1961 for the start of flight training, I was not yet an aviator, so I was relegated to wearing black shoes and boots, except for the white shoes which I was required to wear with summer whites or the Navy dress uniform. Not until I got through flight school and pinned on the gold wings would I be eligible for membership in the Brown Shoe Navy.

Although it may seem rather odd to have such a unique division within one service over something so minor as shoes, the black shoe/brown shoe divide goes back more than a century when naval aviation was just getting off the ground, literally and figuratively.

According to Navy historians, the first naval aviators wore the same low-quarter black boots as everyone else in the service, but when they started aviation training in San Diego, they encountered a particularly dusty airfield that made it almost impossible to keep their boots sufficiently blacked. Pooling what little money they had, the fledgling aviators went to a cobbler's shop in San Diego, where each had a pair of brown boots made. The pilots liked the brown boots because not only did they not show the dust, but they were also more in keeping with the khaki flight suits they wore at the time. After several years of petitions to Navy brass, brown shoes and brown boots were adopted in 1913 as the signature footwear of naval aviators.

The tradition carried on through World War I, World War II, Korea, and Vietnam. Then, in 1976, Chief of Naval Operations (CNO) Admiral Elmo Zumwalt, Jr., decided to end the great shoe divide and by fiat brought aviators into the Black Shoe Navy. The admiral, of course, was part of the Black Shoe Navy and apparently thought it best that everyone in the Navy wore the same color footwear. I would have thought he had more important things to worry about, but by that time, my Navy career was over, so it really had no impact on me. I still consider myself a proud veteran of the Brown Shoe Navy.

Brown shoes returned to the aviation branch nearly a decade later, when Secretary of the Navy John Lehman decreed that the

Brown Shoe Navy was being resurrected at the request of a number of naval aviators. It did not hurt the petitioners that Lehman had been a Naval Flight Officer (NFO), having served as a bombardier/navigator on an A-6 Intruder. An NFO is different from a naval aviator in that NFO's are not actually pilots; instead, they are specialists in weapons and navigation who fly in the back seat of two-seat aircraft.

NAS Pensacola is not the place where naval aviation began, but it has been its home for some time. It is about as idyllic a place as anyone could have found for naval aviation training. Not only is the weather balmy much of the time because of the breezes off the Gulf of Mexico, but the white sand beaches and countless bars along the beachfront made our nights on liberty a relatively inexpensive way to relieve the stresses of training. And, for those of us who had spent the last four years at the Academy, it was a welcome taste of freedom after the restrictive environment of Annapolis. We were not yet naval aviators, but we were on our way to getting those gold wings, and we were feeling our oats.

It was not unusual for us to leave the base at the end of the training day and head to the beach bars in Pensacola. We would be out until 4 or 5 in the morning and then rush back to base just in time to shower and shave before hitting the obstacle course at 7 a.m. With that obstacle course you would have thought we were at a Marine base and not a Navy base, but because we were young, we were none the worse for wear after a night on the town followed by an early morning run through that obstacle course. Such a life is a young man's game, though, and it's doubtful that anyone much over the age of 30 could have handled it.

The real business at hand once we arrived at Pensacola was determining who among us would be jet pilots, who would be prop pilots, who would be helicopter pilots, and who would not be pilots at all but would end up in the Black Shoe Navy as surface warfare officers. Those who flunked out of flight school were said to have "not adapted" and were given other assignments.

Before we even got to the point where we could think about getting into the cockpit of an airplane, we went through several weeks of pre-flight training that included classes in air navigation, flight rules and regulations, aerodynamics, aircraft engines and systems, and meteorology. In addition to the classroom work, there were classes in parachute training, land survival, and water survival.

For those who had issues with the water, or who panicked easily, the water survival portion of training was one of the toughest tests. For some, it was tougher psychologically than it was physically because it involved battling with the infamous Dilbert Dunker, a device well known to generations of naval aviators but not to the general public until the Richard Gere movie *An Officer and a Gentleman* was released in 1982.

The Dilbert Dunker was designed during World War II by a Navy ensign by the name of Wilfred Kaneb. Originally known as the "Underwater Cockpit Escape Device," it garnered the name Dilbert Dunker after a cartoon character of that era known as Dilbert Groundloop, a rather slow-witted soul featured in Navy training videos. Dilbert was the proverbial screw-up: Give Dilbert something to do and he did it wrong.

I suppose the name Dilbert Dunker stuck because it is far more alliterative than the Wilfred Dunker or the Kaneb Dunker.

The Dilbert Dunker was a mockup of an aircraft cockpit set on rails about 10 feet above a deep-water indoor pool. A student wearing a flight suit and helmet would be strapped into the Dilbert as if in a cockpit. The device then was released, and it hurtled down the rails at a 45-degree angle into the water, reaching about

25 miles per hour before splashdown. The Dilbert then turned turtle, leaving the student strapped in upside down in the water until he unlatched his harness and swam out. A safety diver was on hand to assist those who struggled, and more than a few of my classmates came up spluttering and blubbering. Those who failed kept going back until they either succeeded in freeing themselves or gave up and washed out of training.

The whole purpose of the Dilbert Dunker was to teach us how to escape an airplane if we had to ditch at sea and the plane became inverted. Some referred to the training as "teaching us what it feels like to drown." At the time, I just hoped I would never be forced to put that training to use.

At the end of pre-flight training, we were sent to NAS Saufley Field, about 10 miles north of NAS Pensacola, for our introduction to what at that time was the primary training aircraft for the Navy and Marine Corps, the T-34B Mentor. The Mentor was a two-seat, single-engine prop plane with a tricycle landing gear developed by the Beech Aircraft Corporation. It had a top speed of 214 knots (about 250 miles per hour) and a cruising speed of 165 knots (about 190 miles per hour). It was not the nimblest aircraft around, but it served its purpose for me and thousands of other Navy and Marine Corps aviators over the years.

I can't say that I was at the top of my class when it came to flying, but it was something I had dreamed about for so long that my focus and determination made it feel as if flying aircraft was something I had been destined to do. I learned quickly, and, with the help of my instructor pilot, seldom repeated mistakes.

The first flight as student pilots was referred to as PSO-1, or Pre-Solo 1. The primary purpose of the twelve pre-solo flights is to give students a hands-on experience of doing all those things we had been taught and had discussed in the classroom.

First, we determined which plane we were going to fly that day and picked up its "yellow sheet" that provided a recent history of the aircraft which included who flew it and when and whether there were any problems associated with those flights or the plane. After that, we did a pre-flight walk-around of the plane, looking for any cracks or holes in the wings, loose bolts or nuts, how easily the flaps and ailerons moved, and whether the landing gear and engine were in good shape.

Each time I did this with an instructor, he expected me to take more and more responsibility for the pre-flight checks. He followed closely behind to ensure I did not miss anything, but by the time I was ready to solo after a dozen dual flights, the pre-flight checks had become almost automatic. Those pre-solo flights were supposed to be Flying 101: all the basics and few of the frills. I was required to take off and land smoothly, preferably touching down on the first third of the runway. I was taught to recognize stall conditions and how to recover from them.

I was introduced to spins, power-off spirals, steep turns, and low-altitude emergencies, such as flameouts. The textbook answer for a low-altitude flameout was to fly straight ahead and land where possible. One of my fellow students decided he did not want anything to do with that sort of thinking. Instead, when instructed to follow procedures for a low-altitude flameout, he stood on the rudder and banked hard to the left.

"What in the hell are you doing!" the instructor yelled at him.

The student replied, "I'm going to land where I took off." It turns out this student had been a crop duster before he got into the Navy and probably was as familiar with flying as his instructor.

Once we finished PSO-12 and were designated by our instructors as "Safe for Solo," the day finally came when I received the go-ahead to fly the airplane by myself. It was Monday, October

30, 1961, a bright, sunny day with a few broken clouds overhead and the temperature in the mid-70s. It was a perfect day for flying. The instructor climbed into the plane with me but said nothing as I took off from Saufley Field and headed toward an outlying field, where I was told to land.

"Just come back and pick me up," he said as he climbed out of the cockpit.

I took off smoothly and flew around for a while as if I had been doing this all my life. The airplane seemed almost like an extension of me. I felt confident and in charge and was enjoying every moment of that first flight. It obviously was the high point in my life up to that time.

By the same token, I was not so confident in myself that I did not realize that at any moment this piece of machinery that I was strapped into could somehow fail or that I could push it to unacceptable limits and that would be the end of me. I was not sure if the brass would be more concerned about losing me or an expensive aircraft, but I wasn't about to take any chances of losing either. *I came to realize very early in my career that as a pilot I had to have a good sense of my limitations and the limitations of the aircraft. If you are a pilot and don't have that sense, you're probably not going to last long.*

The Navy has a statistic that indicates the most dangerous time for a naval aviator is the first 500 hours of flight time: If you're still alive after 500 hours, then you're probably going to be okay. You will obviously push the envelope if you fly; it's just a matter of knowing how far to push it.

After we had soloed for the first time, we then moved into the acrobatics phase of training, still in the Mentor. We learned to do loops, aileron rolls, barrel rolls, and several other maneuvers.

As my class progressed, some students found that while they

had a desire to fly, their bodies were not always in tune with that desire. One of them was my good friend and fellow card player from the Academy, George Mays.

George had been an enlisted sailor before coming to the Academy and was set on becoming a naval aviator. He did well in the pre-flight classes and the pre-solo classes and, if I remember correctly, even got to solo. What George had trouble with was the acrobatics phase of training. Once we started doing acrobatics, George discovered he was prone to airsickness. Nothing he tried could keep him from throwing up whenever we were required to do acrobatics. He had, as they say, "not adapted."

By the time the Navy decided to drop George from the flight program he was a lieutenant junior grade, which is fairly senior as far as junior officers go, so they made him a surface warfare officer in the Black Shoe Navy. His first assignment was on an oiler, which is a Navy resupply vessel that provides fuel, food, ammunition, and other essentials to ships at sea. The captain of the oiler was an aviator learning to command smaller surface vessels before he was assigned command of an aircraft carrier.

One night the oiler was coming out of the harbor in Charleston, South Carolina, and the skipper said to George, "Give me a steer to clear the harbor."

George said, "Aye, Skipper. Come right to 075 degrees."

"Roger," the skipper responded, "Come right to one-two-zero (120 degrees)."

Once the ship had cleared the harbor, the skipper paused for a moment, and then said, "George, come here," and he took George to the plotting board. "Where were we when you gave me that steer?"

George pointed to a spot on the chart and the captain said, "Where does that take us?"

George pointed to the chart again and said, "Up there."

How deep is the water up there?"

"It looks like about 20 feet," George replied.

"How much do we draw?"

"About 32 feet."

"George, do you think we would have made it?"

Despite those early setbacks in his career, George went on to command a destroyer, taught in the Weapons and Systems Engineering Department at the Naval Academy, and retired as a captain.

Discussions about the U. S.'s fledgling space program were always a topic of conversation at the Officers Club ("O Club") while having a few beers after a day of flying. Alan Shepard, a Naval Academy grad, had flown into space in May 1961, shortly before our graduation and our class crest displayed a rocket ship, so it was little wonder we considered the possibility of one day joining that elite fraternity. I'm not sure whose idea it was, but to test our ability to withstand centrifugal force — one of the prerequisites for space flight —we drove to the nearest coin laundry one night and using our ample supply of quarters, took turns flipping around inside an industrial dryer, which — at least in our minds — closely mirrored the actual NASA test. Each of us escaped with only minor bruises — but none of us became an astronaut.

After the acrobatics phase of training, we then moved into learning to fly in formation, and after that, we learned to fly via instruments. Most pilots I know prefer flying under Visual Flight Rules, or VFR, because you can actually see what's around you. Instrument Flight Rules, or IFR, can be a bit tricky at times, if not dangerous, because the pilot is totally dependent on what the plane's instruments are telling him. Despite their best intentions, those instruments can sometimes lie at the worst time, as I came to discover a few years later in an incident that almost cost me my life and that of another pilot flying with me. Before I arrived at that life-threatening situation, though, I had much more training ahead of me.

Once training in the T-34B was completed at Saufley Field, the assignments for the next phase were made. Those assignments were based on how well we did in the classroom and in pre-solo training, and how well we performed with the acrobatics phase, the formation phase, and the instruments phase. You either got jets, props, or helicopters.

Those picked for helicopters went to nearby Ellyson Field for their introduction to those aircraft that seem to defy all the normal laws of aerodynamics. I was just happy my name was not on that list. As I mentioned earlier, helicopters are an accident looking for a place to happen.

I've flown helicopters and I enjoyed it, but I would not want to do that in combat or for a living. I was a co-pilot when I flew in those helicopters—never in command—because I was not rated for helicopters. Years later, in my job with the FAA I flew a number of different helicopter projects which I probably should not have since I was never helicopter-qualified. I'm just thankful I survived to tell about it because nothing you do in a helicopter is instinctive.

If you were selected for prop planes, you went to NAS Whiting Field, about 30 miles northeast of Pensacola. If you got orders to NAS Meridian in eastern Mississippi, it meant you had been selected to fly jets. My orders read "NAS Meridian," and that meant I was on my way to becoming a jet pilot and fulfilling that dream of being a naval aviator flying onto and off aircraft carriers.

My class arrived at Meridian in late November 1961 and started all over again with pre-flight training specifically related to jet aircraft. There were classes in aerodynamics, classes in jet engines and their capabilities, rules of flight for jets, and all those mundane things you need to know before you actually climb into the cockpit of a jet for the first time.

That pre-flight indoctrination lasted about a month before we actually got to fly what at that time was the standard Navy jet trainer, the T-2 Buckeye. Manufactured by North American Aviation in 1958 specifically for use as a basic Navy training jet, the T-2 was slow, as jets go. Its maximum speed was only 453 knots (521 miles per hour) and while it was a nice-flying airplane, you could not have gotten to Mars in a lifetime in that jet. Some jets just look slow, even when they are flying; the T-2, with its distended underbelly, looked slow and pregnant, whether it was flying or sitting on the ground.

Still, the T-2 served its purpose and introduced us to the intricacies of flying jets, which actually is less complicated than flying a prop plane. With jets, there are fewer things in the cockpit to monkey with than there are with prop planes. With a jet, you don't have to worry about the pitch of the propellers or regulating the air-fuel mixture. All you have on a jet trainer are a few gauges, a throttle, and a speed brake.

One of the highlights of our stay in Meridian, which does not have much in the way of night life for young, rambunctious pilots-in-training, was the wedding of my friend from the Academy, Denny Moore.

Denny asked me to walk his wife-to-be, Kim, down the aisle, since her parents were unable to attend the wedding. I agreed and on the day of the wedding, as Kim and I were waiting in the vestibule, she turned her head to me and whispered, "Dave, I don't know if I want to do this."

I was a little surprised but looked straight ahead and saw Denny standing at the altar and whispered to Kim, "Well, then let's get the hell out of here!"

"You're a big help," she said, but we started walking down the aisle and eventually she and Denny were married, and they have been married for more than 50 years.

Denny went on to fly combat missions in Vietnam before leaving the Navy in 1967. From there he went into the aerospace industry and in 1992 became chairman and chief executive officer of St. Louis-based defense contractor Esco Electronics Corporation, a spinoff of Emerson Electric. He retired in 2003.

After our basic jet training at Meridian, we returned to Pensacola in June 1962 for carrier qualifications, or CarQuals (pronounced "care-kwals"). Also known as CQ, this is the next step in the process of learning to fly jets onto the pitching deck of an aircraft carrier at sea. Pilots call these landings "hitting the boat" while non-aviators call them controlled crashes. There is no easy, slow approach to a carrier deck with a smooth touch-down and long runout to the end of the runway. On a carrier landing you are working with only about 500 feet of deck. If you don't fly the plane onto the deck and hook one of the three or four—depending on the carrier—arresting wires with the aircraft's tailhook, you either drop off the bow of the boat into the water or hit the gas, go around, and try again. Missing all the wires and going around for another try is known as "a bolter," something that happens frequently at night or in rough seas.

The initial stages of carrier landing training are done on land—that way, you don't have to worry about going into the water if you screw up. Known as FMLP (Field Mirror Landing Practice), the object is to get pilots used to "flying the ball" or "the meatball" as it is sometimes known. "The ball" is what pilots call the Fresnel lens, an ingenious gyro-stabilized mirror optical guidance system that helps guide pilots onto the correct glide path for landing on an aircraft carrier.

The Fresnel lens used on carriers is located near the stern (rear) on the port (left) side. A series of amber and red lights are arrayed in a vertical column. The lights are always on but the angle of the lens from the pilot's point of view determines what color he sees and where it is located in reference to a horizontal bar of green lights.

If the amber ball appears in the middle, the pilot is on the proper glide path. If it appears above the horizontal, he is too high. If the light is below the horizontal and turns red, the pilot is too low and runs the risk of crashing into the stern. If too high or too low, the Landing Signal Officer (LSO) standing near the ball will likely wave off the pilot for a go-around and another attempt. Before the invention of the Fresnel lens, carrier pilots had to rely on the LSO with a paddle in each hand to direct them onto the boat. The paddles were difficult to see unless the pilot had extremely good eyesight, so he often did not know if he was on the proper glide path until he was just a few hundred feet from the boat. Although the paddles are long gone, the LSO is still referred to as "Paddles" in person and on the radio.

Training on the ground for carrier landings is constant repetition. An area is marked on the airstrip about the length of the landing area on the deck of a carrier, about 500 feet. You make your approach to the field as if you were approaching the carrier. You find the ball, call the ball to the LSO, and come in on a fairly steep glideslope. You hit the ground hard time after time after time as you would on a carrier and hit the gas for another try. There are no arresting wires during those practice landings, so you leave the plane's tailhook up.

Those landings have been described as sitting in a hard chair and being dropped six feet onto concrete, and a pilot in training does eight or nine of them a day over a two-week period. By the time you head out to the boat, you've done close to 100 of those dry-land practice landings.

At the time I was at NAS Pensacola for carrier qualification in early 1962, the boat was the *USS Antietam*, CV-36. A mere 888 feet long with two forward facing catapults, the angled landing area on the boat was just under 500 feet.

When the day came for us to finally hit the boat, we took off in a four-plane flight formation and headed south to where the *Antietam* was cruising in the blue-green waters of the Gulf of Mexico. As we flew down the starboard (right) side of the carrier at about 1,000 feet, I remember looking down and saying to myself, "Holy shit! Am I going to land on that thing?" It looked only slightly larger than a postage stamp floating on the water.

We banked left and came around down the port side of the carrier, then began to brake. Every 30 seconds one plane peeled off and headed for the boat, giving us a separation of about a minute. The first two CarQuals were touch-and-go. The tailhook stayed up, we spotted the ball, followed it in, hit the deck, hit the gas, and then climbed back into the pattern for another attempt.

Landing speed on an aircraft carrier is about 150 miles per hour. But, at the same time the carrier is moving forward at about 30 miles per hour, creating a good deal of wind across the deck. Since the landing area is angled to the port side, a pilot must account for the crosswind as well. It's a tricky business but once I understood the concept, I found carrier landings to be easier than departures.

When you get catapulted off a carrier, you really have no control for a brief time. The instruments are bouncing all over the place and you can't tell what's going on. It's like a slingshot: once that catapult is released, there's no stopping. You can't put the brakes on and come back for a do-over. You're off the bow of the boat and either you're flying or taking an expensive bath with one of Uncle Sam's multi-million-dollar toys.

The general rule of thumb on the older ships that had only two catapults (cats) down the bow, was that when you launch from the right cat, you make a short turn to the right. When you launch from the left cat, you make a turn to the left. There are safety reasons for that, the most obvious being that if you go into the water, you don't want the carrier running over you.

We had several arrested landings (catching the wire) and several cat shots that first day before heading back to Pensacola. My buddies and I headed for the O Club where, in a well-oiled debriefing session, we discussed our exciting first day of CarQuals. Before I realized it, the time was past 5 p.m., and I was still in my khakis, which was against Navy protocol that dictated "summer whites" in the O Club after 5 p.m.

Also in the bar at the O Club that day were members of the Navy League, a civilian nonprofit organization that supports America's sea services. The Navy Leaguers were in Pensacola to watch carrier landings. A league member from Kentucky cornered me in the bar and began asking questions about my experiences that day aboard the *Antietam*. As the conversation went on, the Chief of Naval Air Training, a three-star admiral, entered the bar and spied me, quickly observing that I was not in the proper uniform. He walked up to me and the ensuing conversation went something like this:

Gentleman from Kentucky: "Admiral, I'd like to introduce you to Ensign West. He's been out on the boat today."

Admiral: "Ensign West, were you in CarQuals today?"

Me: "Yes sir, Admiral." At that point, the admiral noted not only my uniform but also that I was somewhat under the influence.

Admiral, calling over his driver, a Marine Corps enlisted man: "Corporal, please take Ensign West back to the BOQ so he'll be ready for CarQuals tomorrow morning."

End of conversation. I dutifully followed the driver out the door and did not look back. I did not want to see the look on the admiral's face.

Despite that embarrassing encounter with the admiral, I had a great sense of satisfaction with what I had done that day: I felt good about myself and my abilities as a pilot. I still was not officially a naval aviator but was well on my way.

The next stop for me was NAS Kingsville in south-central Texas for advanced jet training and high-altitude gunnery. At Kingsville, we once again went through several weeks of ground school prior to getting into an airplane. We learned first about the F-9 Cougar, and later the F-11 Tiger. Both were manufactured by Grumman and were much faster and much more nimble than the T-2 we had flown for carrier qualifications. The F-9 had a maximum speed of 654 miles per hour, while the F-11 could push just past Mach 1, the speed of sound, at 727 miles per hour when fully loaded. These jets definitely were a step up from what we had been flying.

After ground school, there was an indoctrination phase in the aircraft with an instructor in the back seat before we were allowed to solo. After that came instrument training, followed by the radio phase, where we learned to put navigation information into the aircraft's system. From there we went on to formation flying.

At one point during the training, we were required to plan an instruments-only flight from Kingsville to various spots on the map before returning to base about 90 minutes later. Known to naval aviators as "flying under the hood," this exercise was designed to test flying skills using only the instruments on the aircraft. If a student did not get it right, he might find himself cruising over the Gulf of Mexico or heading to Canada. Usually, everyone is somewhat off in their calculations because the winds at 35,000 feet must also be considered.

To ensure we did not cheat, the instructor put up the hood, sometimes referred to as "the bag," which is little more than a curtain obscuring everything but the instruments on the control panel. It was like flying in an unlit closet with only a few dim gauges to guide you.

When my turn came to make this flight, the instructor cut off the gyroscope as soon as we were airborne. The gyro indicates the aircraft's attitude, or whether it is nose up or nose down. All I had to go on at that point was my heading and what pilots call "needle, ball, and airspeed." The needle and ball give you an indication of whether you are banking left or right or whether the plane's attitude is level. The airspeed indicator helps you figure out when to make your turns based on what you have plotted on your map.

I was flying along, making what I believed were all the necessary turns based on my plot and when we got about 90 minutes into the flight, the instructor said to me, "Mr. West, tell me when we're over Kingsville."

"Okay," I replied and then waited a minute before telling him, "We're over Kingsville right now."

"Pop the hood," he commanded and so I did.

He looked out the window and exclaimed, "You lucky sonofabitch!"

He rolled the airplane over and the Kingsville airfield was right below us.

"You couldn't do that again if you tried," he muttered.

He might have been right. What I had done was a combination of luck and skill because the winds can really be bad at 35,000 feet. The weather information also must have been pretty accurate that day. And, I must have flown my plan fairly accurately to come out right over the field. That experience told me I seemed to have a good feel for this flying business. Air-to-air gunnery practice was done in the F-11 using its four 20-millimeter cannons, each of which could hold 125 rounds. We did not shoot at other aircraft, as we might have to do in combat. Rather, our target, known as "the squirrel cage," was towed at some distance behind another plane. We were in single-seat aircraft at that point but had instructors in the air in their own planes watching us and providing feedback over our headsets. (If you have seen the movie *Top Gun*, you will recognize this scenario.)

High-altitude gunnery is not a skill that is easily mastered. It takes a lot of practice to get halfway proficient and is a skill that has a short shelf life; without continual practice, our skills become rusty.

One of our instructors for that portion of the training was a Marine captain who also was a Naval Academy graduate. His call sign was "Falcon." After the planes were armed and we were sitting waiting to take off, he would come on the radio and say, "Check in, Falcon flight."

We would radio back in order, "Falcon One," "Falcon Two," "Falcon Three," and "Falcon Four."

And he would come up and say: "Mother Falllll...con," which sounded a little like "Mother F...er."

At one point during our training, two of the instructors had a minor mid-air collision. Neither plane was badly damaged, and the only thing really hurt were their egos because they were supposed to be the guys teaching us how to fly air-to-air combat. The students got a big kick out of the mishap, though, and before graduation we had a party and put on a skit that replayed the midair collision. The "perch" is the location where planes wait to perform their gun runs, so in the skit we had one instructor saying to the other just before the crash, "Get off your perch or I'll knock you off!"

The only other notable event during my time in Kingsville occurred when one of my classmates had a bit too much to drink one night and drove his Volkswagen up the front steps of a bar and through the front doors. He took off his cap (also known as a "piss cutter"), stepped out of the car, and walked to the bar as if what he had done was the most natural thing in the world. At least he had the good sense to take off his cap before he got out of the car.

I received my gold wings on December 7, 1962, at Kingsville, the 21st anniversary of the attack on Pearl Harbor in which Japanese aviators flying off aircraft carriers killed more than 2,300 Americans and seriously damaged the U. S. Pacific Fleet. Because it was such a solemn anniversary, there was not the usual revelry following the awarding of the wings.

Now, officially I was a naval aviator and a member of the Brown Shoe Navy. I was eagerly looking forward to my first assignment with a fleet squadron, but the Navy had other plans for me and several others in my class.

Someone, somewhere obviously saw something brewing in the Far East, Vietnam to be more specific. I had gotten indications of it when I was at the Academy, but now that realization of what America's military was about to head into became much clearer. A decision was made to keep seasoned pilots with the fleet in the event they were needed to fly combat missions rather than sending them home to train young pilots. To fill the void, the Navy took some of us who had just gotten our wings and assigned us as instructors.

In January 1963, I was sent back to Meridian as an instructor in the T-2A; obviously, at the time I would rather have gone to a fleet squadron. Looking back on it, though, that time I spent training other pilots proved to be an excellent experience for me. It gave me a few more months watching other dummies doing the same stupid things I had done as a student pilot: It's kind of like sitting in the passenger seat of a car while someone else is driving. You get to see all the mistakes they are making, and you say to yourself, "What the heck are you doing?"

In those days, Mississippi was a dry state—no alcoholic beverages were sold within the state. Not so at the NAS, with regular Navy flights bringing in booze from Florida. The Lauderdale County sheriff got wind of the activity and decided he was going to shut down the operation. He made an appointment with the base Provost Marshal and ordered him to cease and desist. The Provost Marshal replied, "This is U. S. government property, and you have no jurisdiction here. If you don't leave right now, I'll have you thrown in the brig." End of discussion.

In the 10 months I was at Meridian training pilots, I logged more than 700 hours of flying time. Students were required to make weekend cross-country trips with an instructor to give the students experience in interfacing with civilian air traffic control and going in and out of major cities where there was a lot of air traffic. Since I was a bachelor and few of the married instructors wanted to make those weekend trips, I signed on for them and picked up a lot of flight hours that way.

We went to Las Vegas or San Francisco or San Diego or some other destination that was fun. If one of the students lived near a Navy base, we would fly there so he could visit his family. I had a lot of friends in Jacksonville, Florida, so we frequently went there. I usually had two flights a day as an instructor during the week and on the weekends, I picked up 12 to 15 more hours on those cross-country training flights. Sending me back to Meridian as an instructor was probably the best thing the Navy could have done for me at the time because of the wealth of experience it gave me.

As with almost every instance in those years, frequenting the O Club was *de rigueur* after a day of flying. One guy who often accompanied me was Sid Cruise, a Navy dentist. Sid was a graduate of Vanderbilt University and the School of Dentistry at the University of Louisville and quite a partier! One night he burned his lips while trying to down a Flaming Hooker, which is a shot of Drambuie set on fire. Of course, the idea is to down the liquor quickly and leave the flame in the bottom of the shot glass, but Sid was unsuccessful. I imagine his CO gave Sid a curious look the next morning when he reported for duty!

I was at Meridian as a trainer until December 1963, when I was reassigned to an East Coast air replacement wing based at Key West. I knew from there I eventually would end up with a fleet squadron. First, though, I had to go to Norfolk, Virginia, for a night instrument refresher flight in an A-4 trainer.

The instructor this time was the executive officer (XO) of one of the squadrons at Norfolk. Just as the instructor had done at Kingsville, when I did the night instrument flight test, the XO put up the hood as soon as I was airborne and took away the gyro, leaving me with only the needle, ball, and air speed.

I was sitting there in the cockpit fat, dumb, and happy and everything looked fine; I was thinking that this flight was going to be another test like the one at Kingsville and would be a breeze. All of a sudden, the altimeter started unwinding and the airplane started to roll. I looked at the instruments for a clue, but everything looked normal.

Despite what the instruments were telling me, I put in some

left stick and pulled back to gain altitude. As soon as I regained altitude and air speed, it happened again. The altimeter was unwinding, and the plane was rolling. I looked at the instruments and they indicated I was going straight and level. I could tell that was not the case because the plane was going down at a rather rapid pace. I could see it on the altimeter and the air speed indicator.

"Are you all right?" the XO asked.

"I don't know what's going on," I said, "but this needle is perfectly still and I'm turning."

"Let's go home," he said, "and we'll talk about it."

The next morning, I got to the office and was told the XO wanted to see me. I stepped into his office, and he looked up from his desk and said, "You are one lucky sonofabitch."

"Why is that, sir?" I asked.

"We found out the needle ball was frozen. It was showing that you were straight and level when in fact you weren't. That was an instrument failure you had, and how you recovered without any instruments, I'll never know."

"Well, sir," I said, "I was turning left so figured I'd better turn right, and I was sinking, so I figured I'd better climb."

"Well, those two were the right things to do," he said, shaking his head. "In all my years of flying, I have never seen that happen."

Although that flight could have ended disastrously for both of us, it taught me three things: First, it taught me that instruments will fail. Second, it taught me to rely on my instincts if I sensed there was something wrong with the instruments. Third, and most importantly, it taught me that a good pilot straps the aircraft on his back. In other words, a good pilot controls the aircraft, not the other way around.

So, that was twice in a very short time that I was a lucky sonofabitch, if for no other reason than I trusted my instincts as a pilot. It would not be the last time listening to my gut instincts about airplanes and flying got me out of trouble.

CHAPTER 4
'ROGER BALL' AND THE
TOPHATTERS

At the end of the 10 months I spent training young pilots at NAS Meridian, I was sent to Fighter Squadron 101 (VF-101) at NAS Key West in December 1963. Known as the "Grim Reapers," VF-101 was a Fleet Replacement Squadron (FRS) at that time. Pilots were sent there to check out in the aircraft and await an assignment to an operational squadron in one of the two East Coast fleets.

Despite the tropical setting and easy-going lifestyle of the city, NAS Key West was not exactly a place to relax and take it easy. Sometimes called the "Gibraltar of the Gulf," the base was still teeming with Army, Navy, and Air Force activity a little more than a year removed from the Cuban Missile Crisis of 1962. The Cold War tensions in the Caribbean remained high, and since Key West is only 90 miles from Cuba, there were concerns that the Soviet Union might be planning something new in Cuba after being forced to withdraw its long-range missiles.

Although VF-101 was a replacement squadron, pilots had to maintain a high standard of readiness because of our location near Cuba and the fact that any of us could be called up to a fleet squadron at any time. The squadron had only recently transitioned from the F4D-1 Skyray and F3H-2 Demon to the larger and more powerful F-4H-1 Phantom II, later simply referred to as the F-4B Phantom. Manufactured by McDonnell Douglas, the F-4 was designed for multiple missions for both the Navy and the Air Force. It was supposed to be an interceptor, a fighter-bomber, and an aerial reconnaissance aircraft. In Vietnam, it also was used for close air support (CAS) for ground troops.

Ask 50 pilots who flew the F-4 in combat and during peacetime, and I am sure the large majority would say it was a good airplane. And, it was a good airplane: it had a lot of power, a lot of speed, and could hit Mach 2.4.

The plane's other qualities weren't as good, however. The problem was that it had been designed to do too many missions. You can't design an airplane to do every mission that every service wants it to do. It's like trying to build the perfect animal and ending up with a giraffe or a duck-billed platypus. An aircraft can't be a good fighter, a good attack aircraft, and a good bomber all at the same time; one of those missions, if not all, will suffer to some degree.

Who drives the idea that an aircraft can be designed to do multiple missions for multiple services and do them all well, I'm not sure. The two likely candidates are the manufacturers and the U.S. government. One other case with which I am familiar is the T-38, which was an Air Force trainer. Northrop built that and sold it to the Air Force. The Air Force had very little input on the design of that plane. But, on many other airplanes the military services put out a request for proposal (RFP) for a particular type of aircraft, and three or four companies bid on it. Then the government picks the aircraft it believes will perform the mission stated in the

RFP. Sometimes the government is right, and sometimes it's dead wrong.

For me, the only thing the F-4 proved was that if you put enough thrust on a barn door you could probably fly it. The F-4 had 34,000 pounds of thrust, but flying it was not unlike flying a barn door. It was not that I did not enjoy flying that airplane, though. I had a wonderful time with it. To stay proficient in the aircraft, I frequently flew from Key West to Jacksonville, Florida, to visit friends there and the girl who a year later would become my first wife. Several single pilots lived off base in a house we called "The Snake Ranch," which frequently was the scene of parties at which a number of young women were present. Kathy was at one of those parties when I flew in from Meridian earlier in the year, and I continued to see her after I was transferred to Key West. We were married in 1964 before my first Mediterranean cruise.

Returning to Key West from one of those flights to Jacksonville, I decided to see just what that F-4 could do. In those days, you were supposed to inform air traffic control (ATC) if you changed your air speed by more than five miles per hour. (That rule struck me as being rather stupid because five miles per hour in a jet is peanuts.) I told the guy in the back seat flying with me, "We're going to go across to Tallahassee and then turn south and increase our air speed."

We cleared Tallahassee, turned south, and I called the air traffic controller and said, "Sir, I'd like to change my air speed."

"What do you want?" he asked.

"Sixteen hundred."

"One six zero (160)?"

"No. One six zero, zero (1,600)."

He didn't un-key the mic and I heard him say to his buddy in the control room, "Hey, Al, watch this."

I hit the afterburner and we got out to about Mach 2.4, and by the time we hit that speed, I would have passed over Key West if I hadn't started descending. That was the fastest barn door I ever flew.

The speed of the F-4, and how quickly it could accelerate, surprised a number of the students I taught to fly the plane. It also cost those students a number of beers.

As a way of introducing a new student to the aircraft's power, I would ask, "How long do you think it will take me to go Mach 1 (the speed of sound)?"

"Maybe a minute," was the usual response.

"Would you believe 20 seconds?"

"Nah, there's no way."

"Want to bet a beer on it?"

"Sure."

I settled into the cockpit with the student in the back seat and climbed to 35,000 feet while accelerating to .9 Mach. Then, I pulled the aircraft straight up and let the speed decrease to just above stall speed. (You have to be gentle; you can't put in the ailerons, or you will stall.) I put in a little rudder and tipped the plane over so the nose was right on the horizon. I lit the burners and said to the student, "Start the clock."

Every time I did it I got to Mach 1 in less than 20 seconds and every time that student had to buy me a beer.

I stayed with VF-101 until May 1964 before receiving orders to a fleet squadron. The original orders were to Fighter Squadron 84 (VF-84), the Jolly Rogers, based at NAS Oceana in Virginia Beach, Virginia. However, before I could make that move, my orders were changed and I was assigned to Fighter Squadron 14 (VF-14), when one of that squadron's pilots was lost in an accident during a Med Cruise on the carrier *U.S.S. Roosevelt*, CV-42.

According to VF-14 history, the unit was performing night operations in July 1964 when one F-4B Phantom had both its engines flame out just as it was launched from the carrier. The RIO (Radar Intercept Officer), the back seater, was able to eject. He landed in the water just as his chute deployed and was rescued. The pilot, Lieutenant Dale N. Fendorf of Kansas City, Kansas, a 1959 Naval Academy graduate, was unable to eject and was lost at sea.

VF-14 is known as the "Fighting Tophatters," or just "Tophatters," call sign Camelot, the oldest continuously operating squadron in the Navy. The squadron's motto is "Oldest and Boldest" because it was formed in 1919 and began its service on aircraft carriers in 1926 as Fighter Plane Squadron One. It got its nickname Tophatters from one of the squadron's pilots who always went on liberty dressed in a top hat and a tuxedo with tails.

I joined the squadron at NAS Cecil Field in Jacksonville, Florida. VF-14 was supposed to be on a cruise in the Mediterranean at the time, but the planes were sent back to Jacksonville when the *Roosevelt* lost a blade on one of its propellers and had to limp back to Bayonne, New Jersey, for repairs.

When I arrived at VF-14, repairs on the *Roosevelt* had been ongoing for some time, and the ship was almost ready to head back to sea and complete that Mediterranean cruise even though there were only about two months left in that rotation.

Going on a Med cruise, or any cruise on an aircraft carrier, can be enjoyable and exciting. There are usually great ports of call for liberty, and there is plenty of time to learn the intricacies of air operations on a carrier. Landing on an aircraft carrier is a perishable skill, and constant practice is needed to maintain proficiency to ensure you don't become a statistic by losing an aircraft or yourself—or both.

For anyone not familiar with air operations on a carrier, the flight deck during those operations can look like chaos in a very small space as dozens of wind-whipped sailors in different-colored shirts move around in a seemingly random pattern during aircraft take-offs and landings. Despite what it may look like, every movement and every hand signal is part of a carefully choreographed ballet designed to get aircraft off the boat, into the air, and back onto the boat safely. Each color shirt indicates a specific responsibility for the crew member wearing it, and anyone who has ever served on a carrier knows exactly what that responsibility is.

The yellow shirts are assigned to officers of the various crews on deck. These include aircraft handling officers, plane directors, catapult officers, and arresting gear officers. Aircraft handling officers are responsible for how the aircraft are arranged on the deck. Aircraft directors are responsible for directing all aircraft movement on the flight deck and the hangar deck below. Catapult officers, also known as "shooters," make sure there is enough steam built up in the catapults to launch aircraft and then give the pilot the signal to launch.

Red shirts handle ordnance and fire-fighting duties. Purple shirts are gas passers, handling fuel for the aircraft. Brown shirts are air wing plane captains who get the planes ready to fly. Green shirts are responsible for the catapults and arresting gear. Blue shirts chain and chock aircraft at the direction of the yellow shirts.

Among those who wear the white shirts are the landing signal officers, or Paddles. As mentioned earlier, LSO's stand near the Fresnel lens on the port side of the boat and are in radio communication with pilots as they come in for a landing. A good LSO can be a real lifesaver. That's especially true when the boat hits rough seas. The Fresnel lens almost becomes a liability when the ball is going up and down so much that it is virtually impossible to tell where the deck will be when you are trying to land a 40,000-pound aircraft traveling at 150 miles per hour on a 500-foot airstrip.

A good LSO can position a fake meatball on a mirror that a pilot focuses on. If he puts the meatball up, it means you're too high and you come down. But, if he puts the meatball down, it means you're too low and you go up. Flying onto an aircraft carrier takes a lot of faith that the LSO knows what he is doing because he is estimating where the boat is going to be when you get there. Once you become an LSO, and you're good at it, word gets around and that's the job you get, whatever boat you go to.

I knew of a case where the LSO and assistant LSO were standing on their platform during flight operations and a plane hit the stern of the carrier as it tried to land. They dove for the protective net on the port side, but one of them missed the net and went into the water. It was early evening, and the rescue team didn't find him until the next morning. After being rescued, he said there were a number of times during the night when he thought that it was the end and that he was going to die, so he held his breath, stopped paddling, and just let himself sink. After about 15 seconds, he'd paddle back to the surface and say to himself, "Well, I think I'll give them just a little more time."

During daylight operations, the Fresnel lens becomes visible once the pilot makes his final turn at about three-quarters of a nautical mile to approach the boat. At that point, the LSO will radio him and say, "Call the ball."

The pilot will respond with his squadron's call sign, aircraft side number, aircraft type, "ball," and fuel state. On my flights in the F-4, after I made that final turn and saw the ball, my call might have been something like, "Camelot. One Zero Two. Phantom. Ball. 3.5." Camelot was the VF-14 call sign (I was Camelot-8 because I was the eighth most senior pilot in the squadron), 102 was the aircraft side number, Phantom was the type aircraft, Ball, and 3.5, meaning I had 3,500 pounds of fuel left.

Once the LSO heard my report, he would respond, "Roger Ball." Over the years, "Roger Ball" has become the name of a prototypical, albeit mythical, naval aviator whom everyone has heard of, but no one has ever met.

After the "Roger Ball" call, the pilot would not hear anything else in his headset unless the LSO required minor corrections to the glide path or if he waved off the pilot because he was too high or too low. If all went well, the plane's tailhook would hit the deck between the second and third of four wires spaced about 50 feet apart and snag one of them. The goal is to catch the third wire on a consistent basis. The wires are woven from high-tensile steel wire and are attached below the deck to hydraulic cylinders that are able to bring the aircraft's speed from about 150 miles per hour to zero in about two seconds.

As soon as the aircraft hits the deck, the pilot applies full power so that he has enough speed to take off and get back into the pattern for another try in case the tailhook fails to snag any of the wires, "a bolter." Every landing on a carrier is fraught with danger, no matter how calm the seas are, and after every successful trap, the adrenaline rush that accompanies the landing takes a while to wash out of your system.

There was some concern among the brass in those years that the constant G-forces to which jet pilots were subjected made us more susceptible to hemorrhoids. This was especially true for carrier-based jet pilots because not only were we frequently fighting G-forces in the air, our butts were also constantly being battered by the controlled crash landings we had to make every time we landed on the boat. No wonder carrier pilots were sometimes referred to by other pilots as real hard asses.

In an effort to get some sense about whether hemorrhoids were an actual concern for carrier pilots, our flight surgeon and my buddy, Howie Berg, did monthly checks of the health of the pilots, which usually included a mass public examination for hemorrhoids. Howie would call all the pilots onto the flight deck on a day on which there were no air operations and yell out, "Drop your pants! Bend over and spread your cheeks!"

Howie would then go down the line checking bare bottoms for hemorrhoids. At least that is what he said he was doing. One day one of the pilots in my group remained standing and when Howie approached him, he opened his mouth and spread his other cheeks. Howie slapped him on the butt and barked out, "You're a real *smart* ass!"

Of course, this was at a time when carrier crews were all male. You couldn't get away with a stunt like that today since women have become an integral part of all shipboard crews in the Navy.

As frequently as we worked on cat shots and traps while at sea, there also is a significant amount of down time on board a ship, especially for pilots if the weather is bad or if there are mechanical issues that prevent flight operations. In the pre-computer, pre-Internet era, I spent a lot of time playing cards — especially bridge, pinochle, and canasta — frequently supplementing my Navy pay with money I won from my shipmates.

Those down times also provided an opportunity for people to tell sea stories (some of which may even be true), with every storyteller trying to outdo the last guy with some tale of something he heard or witnessed — or thought he heard or witnessed.

One of my favorites involved an enlisted sailor and a chief petty officer. A carrier had docked in a port and most of the crew was getting ready to go on liberty. On an aircraft carrier there is a forward brow and an aft brow. The officers arrive and depart from the forward brow and the enlisted from the aft brow. On the forward brow, the officer of the watch is a commissioned officer and on the aft brow, it's usually a chief petty officer. The officers salute the flag and say, "I have permission to go ashore." The enlisted say, "May I have permission to go ashore?"

The story goes that this particular E-6, a petty officer, and the ship's chief petty officer who was the officer of the watch that particular day, did not get along well. When the petty officer asked for permission to go ashore, the chief looked at him and said, "Your shoes aren't shined. Go back and shine your shoes."

So, the petty officer went back and shined his shoes and came back and again asked the chief for permission to go ashore. The chief looked at him and said, "Your tie's not right. Go back and fix your tie."

The petty officer went back again. The third time he stood before the chief he was wearing the Medal of Honor around his neck. He looked at the chief and said, "I *have* permission to go ashore."

Now, whether that story is true or apocryphal, I don't know and there's really no way to prove it, but it is one of the better sea stories I ever heard.

On December 23, 1964, the *Roosevelt* and VF-14 returned from sea just in time for Christmas, and we began the process once more of re-fitting the ship and re-training the crew. The VF-14 skipper at the time was Commander Lionel "Mighty Mike" Ames, Jr., a Naval Academy alumnus and a graduate of the Navy's test pilot school. He was quite emphatic about how he wanted certain things done and how he absolutely did not want certain things done to the point of being obsessive about everything. For example, in the Ready Room we had chairs with space underneath to store our flight gear during briefings. Usually when we went into the Ready Room, we grabbed the first available chair and stuffed our gear under it. Ames decided he wanted a chair all to himself and let everybody know that *nobody* sat in his chair, even when he was not there.

On another occasion, Ames called a meeting of all the pilots, so we all dutifully filed into the Ready Room, making sure not to sit in the skipper's chair. When we were seated, Ames held up a red pencil and said, "There is only one red pencil in this squadron, and this is it. I better not catch any of you using it!" He used that red pencil to establish his authority because if you saw something written in red, you knew it came from Commander Ames.

Ames also liked to schedule himself for the first flight in the morning as part of a four-plane formation. At that time, we were using pneumatic air-start hoses to crank up the two turbojet engines on the F-4, and Ames always wanted all four planes in the flight started at the same time. It was my job as line division officer to make sure the planes were properly maintained, were ready to fly on time, and that all four planes were started at the same time.

Seldom was I able to have all four jets cranked and ready to go at the same time because we did not have enough hoses for all four; the hoses kept breaking, and obtaining replacements was difficult because we were the last F-4 squadron at NAS Jacksonville. The Navy was moving all fighters to Norfolk and moving all attack aircraft to Jacksonville, so our resources were limited.

One day after Ames came back from his flight, I got a note that said, "Lieutenant West, see Commander Ames."

I went to his office and he looked at me and said, "Dave, we didn't get four airplanes started on time this morning."

"Skipper," I replied, "we don't have the hoses."

"Well, get the hoses! I want those airplanes started at the same time!"

I lost it at that point. I said, "Goddammit skipper, if you want me to, I'll sleep in the shack at night and get those airplanes started! I'll do it!"

"Calm down, Dave," he responded with a wry smile. "It's not *that* serious."

I never heard from him again about getting those planes started at the same time.

Although I was not completing regular carrier landings, also referred to as "traps," while the *Roosevelt* was back in Mayport, I kept up my flight proficiency by spending as much time as I could in the F-4. One day I decided to go to Key West to see J. D. Morgan, a flight surgeon who was a friend of mine. J. D. was a good swimmer and fisherman and used to catch a lot of Florida lobsters (also known as rock lobsters). He would catch them, put eight to 10 tails in half-gallon milk containers, fill the containers with water, freeze them, and sell the containers for $2 each, which was quite a bargain even in those days.

One weekend not long after we got back from the Med cruise, my friends and I decided to have a squadron party. I thought it would be a good idea to fly down to Key West, see my friend J. D., pick up a bunch of lobsters, and come back to Jacksonville for a lobster feast. It was unknown to me, but after I had been stationed there with VF-101, the field at Key West had been closed to all air traffic except for official business. My lobster run was not exactly "official business." Nevertheless, I flew to Key West with Howie Berg in the back seat of the F-4. As soon as we landed, this old Navy commander stopped me as I climbed out of the airplane and said, "Lieutenant, what are you doing here?"

More than likely, he was chief of the watch on duty that day.

"I'm here to get some lobster, Sir," I responded innocently.

He put his hands on his hips, looked at me in a not-too-friend-ly way and snarled, "Did you know this field is for official business only? You get some gas, and you get the hell out of here!"

I looked back at the plane and saw one of the engines was leaking oil, so I replied, "Sir, I'm not going anywhere because I have an engine that needs repairs."

"You're going whether the engine's leaking oil or not!" he ordered.

We seemed to be in a bit of a standoff, so I called the commanding officer of the Replacement Air Group, who was senior to this officer, and told him what my problem was.

"I'll be right over," was his response.

So, the RAG officer came over and told the watch officer, "That airplane is not going anywhere. It's grounded."

Howie and I spent the night in Key West, had lobster for dinner, and went back to Jacksonville the next day with more lobster.

By the time the *Roosevelt* was ready for another Med cruise in June 1965, Kathy was pregnant with our first child, and it was obvious I would be at sea when she delivered. My father was especially excited about having grandchildren and every so often before I left on the cruise he called and asked, "How is Junior doing?" even though we were not sure if the baby would be a boy or a girl.

I told him, "Dad, there aren't going to be any Juniors in this house." So, when our first child was born in October 1965, and it was a boy, I purposely named him David Paul West, II, so that nobody would pick up that "Junior" bit, but technically he is a junior.

We left for the Med on June 28, 1965. Just three months earlier, the Marines from Battalion Landing Team 3/9 (3rd Battalion, 9th Marine Regiment), waded ashore at Red Beach Two, just north of DaNang, South Vietnam. They were loaded down with combat gear, full packs and helmets, and M-14 rifles without any ammunition.

Upon hitting the beach, the Marines were greeted by curious onlookers, South Vietnamese Army officers, girls tossing flowers, and a large sign that read, "Welcome Gallant Marines." Although the U.S. had had military advisers in South Vietnam for some time, this was the first indication that America was being drawn inexorably into a war that hardly anyone at home knew much about.

Within the next few years, that initial contingent of 3,500 Marines swelled to more than 500,000 American military personnel in South Vietnam. And, within 18 months those of us in VF-14 would be flying combat missions over North Vietnam in an effort, in the words of former Air Force General Curtis LeMay, "to bomb them back to the Stone Age."

"Roger Ball" would see more action on more aircraft carriers over the next seven-plus years than anyone ever thought possible as naval aviation became an integral part of the air war in both North and South Vietnam. Yet, it seemed not so much a war as it did an effort to produce a settlement without victory. There was an unsettling and frustrating uncertainty from the beginning about what our mission was and how we were supposed to carry it out. It was like playing a game of whack-a-mole with deadly consequences for a number of pilots because the moles were shooting back.

CHAPTER 5
VIETNAM

Where Figures Lied and Liars Figured

No sooner had I returned from my second Med cruise in December 1965 than I learned VF-14 was being reassigned from NAS Cecil Field in Jacksonville, Florida, to NAS Oceana in Virginia Beach. The *Roosevelt*, meanwhile, remained homeported at NAS Mayport in Jacksonville.

And, not long after that move to Oceana in late January 1966, we were told to prepare for another move; this time the *Roosevelt* and its entire air wing were being sent to the U.S. Seventh Fleet, headquartered in Yokosuka, Japan. That meant only one thing: our next assignment would be in the Western Pacific, specifically Southeast Asia. Instead of cruising in the relatively peaceful waters of the Mediterranean or Caribbean, we were bound for the decidedly hostile waters of the Tonkin Gulf off the east coast of North Vietnam.

I knew that as a naval aviator I eventually would end up flying combat missions, most likely over the North; it was just a matter of when. Pilots were in great demand for an air war that was growing in intensity in North Vietnam just as fighting on the ground in South Vietnam was increasing.

The most notable ground battle to that point had been in the Ia Drang Valley of South Vietnam in November 1965. It marked a turning point in the war because it was the first major engagement between main-line troops of the North Vietnamese Army (NVA) and regular U.S. Army forces. Two battalions from the 7th Cavalry Regiment of the U.S. Army's 1st Cavalry Division (Airmobile) took on an estimated 2,000 North Vietnamese regulars in that battle. Over three days of fighting, the Americans suffered 251 killed and 245 wounded while the NVA had an estimated 1,500 killed before retreating.

That battle, which was hailed as an American victory despite the numbers killed and wounded, later became the subject of the best-selling book, *We Were Soldiers Once... and Young* by Harold G. Moore, one of the battalion commanders, and Joe Galloway, a journalist with the unit during the fighting. In retrospect, what that battle also showed was that the war in Southeast Asia would be a war of numbers. It would be a war of enemy body counts and of how many sorties were flown, how many bombs were dropped, how many bridges and buildings were destroyed, how many airplanes were lost, how many American troops were killed or wounded, and how many were taken prisoner.

To paraphrase an old maxim, Vietnam quickly became a war in which figures lied and liars figured. The political "wizards" in Washington had already grabbed control of the war's strategy from the military and decided the best way to bring the North Vietnamese to the bargaining table was to bomb them into submission in what was code-named Operation Rolling Thunder. But, the only way Washington had to measure the war's progress and to try to convince the American public we *were* making progress was with numbers that were often inflated or deflated to fit specific political agendas.

What no one in Washington considered, though, was the resolve of the North Vietnamese to reunite a country that had been split at the 17th Parallel in 1954 after the French defeat at Dien Bien Phu. By early 1966, the North Vietnamese were infiltrating tens of thousands of troops and war materiel into South Vietnam on a regular basis—no matter how much we bombed them.

Of course, I did not know any of this at the time I was ordered to Southeast Asia. It was only after I got there and saw first-hand how the war was being micromanaged by Washington that I became aware of how ill-defined the goals were and how many restrictions were placed on where we could fly missions and what we could and could not bomb. And it was only after an unexpected chat with an admiral who was in command of the air war that I began to understand how much Washington had its hands in both the strategic decisions and quite often the tactical decisions, the latter which are usually best handled by commanders in the field.

In late January and early February 1966, the *Roosevelt* and Carrier Air Wing 1 (CVW-1) participated in a four-week shakedown cruise off Puerto Rico in which we conducted air defense and anti-air-warfare exercises. We lost one aircraft during those exercises when an F-4's engines flamed out. The pilot and his RIO ejected and were quickly rescued.

Not long after that brief deployment, the squadron's pilots and RIO's were sent for several weeks of training at the Navy's Survival, Evasion, Resistance, and Escape (SERE) School. Since we were heading to fight in a tropical climate, it would have been a reasonable assumption that we would be sent to a SERE school with a similar environment. The U.S. Air Force had just such a school at Howard Air Force Base in the Panama Canal Zone. That assumption, however, does not take into account how jealously each service guarded its own piece of turf, and cross-service cooperation was not something to be asked for or given in those days.

Instead of sending us to Panama for SERE training, the Navy sent us to its school in southwestern Maine near the little mountain town of Rangeley. We arrived there toward the end of February 1966. To say it was cold in no way describes just how miserably cold it was. It was Michigan's Upper Peninsula in the winter cold—if not colder. Temperatures usually ranged from a low of minus 20 to a high of plus 10. There was snow, ice, freezing rain, and wind that made things even more miserable since most of SERE school is spent out in the elements. And this was supposed to teach us how to survive in a jungle environment if we got shot down? It made no sense then, and still makes no sense. But that's our government for you.

After about a week of classroom work in which we were instructed in how to live off the land while evading capture if we were shot down, we were sent out into the woods to put that teaching into practice. We were dropped off somewhere in the middle of nowhere with a map and told to find our way to a certain spot. Once we got there, it turned out to be a prisoner of war (POW) camp where our captors threw us into little rooms that were supposed to replicate cells. The change of scenery beat being outdoors, but not by much.

Although we knew this was not the real thing, the instructors tried to make the training as realistic as possible without actually injuring us. Stories were filtering back from Vietnam that pilots shot down in North Vietnam were being tortured by their captors, so we were subjected to a number of strenuous interrogations meant to test our individual resolve and to drum into our heads that if captured we needed to adhere to the U.S. military's Code of Conduct to help get us through. That Code stresses resistance at all costs for American POWs, even to the point of death, and absolutely no collaboration with the enemy.

The trainers did other things to try to keep us off balance like depriving us of sleep and a reasonable amount of food. Somehow, I ended up being the cook for our little band of POWs for the few days we were in that camp, and I remember at one point being given what looked like a beautiful ham. It turned out to be nothing but a bone wrapped up to look like a ham. There was not one piece of meat on that bone.

The training was so realistic for some "prisoners" that they got into fights with the trainers, resulting in a few minor injuries. When things got too heated and there was a risk of serious injury, someone would play a recorded version of the National Anthem, and everyone would stop what they were doing, snap to attention, and render a salute.

Since I had spent a lot of time as a youngster hunting and fishing in northern Michigan, I got through the school without any real problems. However, the one thing I took from that experience was the conviction that I did not *ever* want to be a prisoner. I vowed to my wife at the time, "If I get shot down and you don't hear from me in three months, I'm dead. I'm not going to sit in a prisoner of war camp."

By the time I finished SERE School, some of my fellow naval aviators had already been sitting in a prison in North Vietnam for more than a year. Lieutenant (junior grade) Everett Alvarez, Jr., was shot down and taken prisoner in August 1964, the first naval aviator to suffer that fate. He was not released until 1973 after spending more than eight years in captivity. I had no intention of joining him or any of the other American POWs in a North Vietnamese prison.

We did one more shakedown cruise on the *Roosevelt* in April and May of 1966, this time in the waters off Guantanamo Bay, Cuba, before flying back to Oceana for additional training that included tactical briefings and simulated air strikes involving the entire air wing. Our Phantoms also were equipped at that point with what we were told was the latest in Electronic Countermeasures (ECM) equipment. North Vietnamese radar sites were having some success lighting up our aircraft for Surface to Air Missile (SAM) strikes, and this new ECM was supposed to give us a better chance of defeating them. However, the new equipment proved to be little better that what we already had. According to the squadron's final report to the Chief of Naval Operations following our tour, there was "little or no improvement in operational capability of the F-4 weapons system" due to lack of spare parts and an "extremely 'dirty' EW (Electronic Warfare) environment of North Vietnam."

We finally departed Mayport for Southeast Asia on June 21, 1966, with more than 80 aircraft jammed onto the flight and hangar decks. The following squadrons and the planes they flew were on board as a part of CVW-1:

Fighter Squadron 14 (VF-14) Tophatters (F-4B Phantom II)

Fighter Squadron 32 (VF-32) Swordsmen (F-4 Phantom II)

Attack Squadron 172 (VA-172) Blue Bolts (A-4C Skyhawk)

Attack Squadron 12 (VA-12) Flying Ubangis (A-4E Skyhawk)

Attack Squadron 72 (VA-72) Blue Hawks (A-4E Skyhawk)

Heavy Attack Squadron 10 (VAH-10) Vikings (KA-3B Skywarrior)

Carrier Airborne Early Warning Squadron 12 (VAW-12) Bats (E-1B Tracer)

Light Photographic Squadron 62 (VFP-62) Fighting Photos (RF-8G Crusader)

Helicopter Combat Support Squadron 2 (HC-2) Fleet Angels (UH-2A/B Seasprite)

Once we set sail from Florida, we took a rather meandering route to our destination. Our first stop was St. Thomas, U.S. Virgin Islands, where an Operational Readiness Team went over the ship and the air wing to see if we were actually ready for what was ahead of us. Following that inspection, we sailed down to Rio de Janeiro for three days of shore rest and recuperation (R&R).

From Rio we headed east around the southern tip of Africa and the Cape of Good Hope. Since this was winter in the Southern Hemisphere, during a portion of that passage storms kicked up swells that some estimated at 60 feet or more; there was not much flying done on those days.

At some point during the trip, the Marine Corps detachment on board gave the aircrews an opportunity to do some target practice with the 9-millimeter handguns we were issued. A number of pilots and RIO's preferred a snub-nosed .38 caliber pistol because it was lighter and more compact, but the semi-automatics were available for those who preferred them. The Marines threw some targets off the fantail, and we stood there plinking at the targets.

After one shooting session a tough old Marine gunnery sergeant walked up to me and said, "Lieutenant, if I were you and was about to be captured, I'd throw that damn gun away because your pistol isn't going to hold off 30 Vietnamese with machine guns. So, if you want to live, throw away your pistol."

What he said made sense, but I still had no intention of ever being taken prisoner.

On August 1, 1966, the *Roosevelt* arrived at Subic Bay in the Philippines. While the carrier took on ammunition and fuel for the aircraft plus fuel and supplies for the ship, most of the air wing was sent to NAS Cubi Point for additional training. We departed the Philippines on August 7, 1966, for the Tonkin Gulf and what the Navy referred to as Yankee Station, which was a marshalling area roughly 100 miles offshore the North Vietnamese city of Dong Hoi. Another carrier group was loitering in the waters off South Vietnam's Mekong Delta at Dixie Station, flying missions in the southern part of the country.

The *Roosevelt* was the first carrier to go directly to Yankee Station from the Philippines. All the others would go to Dixie Station first and cruise around for a while, getting their courage up to go north, I guess. Despite our lack of experience, we began combat operations as soon as we got to Yankee Station. The *Roosevelt* and its air wing were just one component of what was known as Carrier Task Force 77 (CTF-77), which usually consisted of three or four aircraft carriers plus various cruisers and destroyers.

During the time we were flying combat missions, we were said to be "on the line." This is where the "numbers war" I referred to earlier came into play. While we were "on the line," numbers were a key component of everything we did. A careful record was kept of how many sorties were flown and by which aircraft, how many bombs were dropped, how many missiles were fired, how many targets we claimed we hit, and how many aircrews were killed, declared missing, or captured.

We usually flew 12 hours on and 12 hours off. Pilots flew either midnight to noon or noon to midnight. We did that for about a week and then would swap off, so we either flew 24 hours straight or had 24 hours off during the switchover.

The missions varied from day to day, especially for those of us flying F-4s. The Phantoms were not as maneuverable as the A-4s, so we often flew reconnaissance missions, or "recces" (pronounced WRECK-ees). These were missions in which we were basically looking for targets of opportunity such as truck convoys or ammunition storage sites.

We were told not to drop bombs on schools or hospitals; those were the best targets, though, because that's where the North Vietnamese kept their ammunition and supplies. They learned quickly that if they painted red crosses on those buildings that were actually military warehouses, a lot of our pilots avoided them. If you wanted to disrupt the troops, though, you bombed those places; the red cross provided a good target, but you never told anyone what you had done after you came back to the ship.

After every mission a debriefing officer would question each pilot about where he dropped his bombs. He'd say, "Lieutenant West, what did you do?"

And I would respond, "What was I supposed to do?"

He'd look at his mission list and say, "You were supposed to recce this area and look for targets of opportunity."

And, I would say, "That's exactly what I did. I recced that area and I got 16 trucks."

Then he would write down the number and send it up the line to add to our total numbers. They had no way of checking whether you actually did what you were supposed to do (or what you told them you did) and in many cases the numbers reported to higher command were greatly inflated.

We usually carried a combination of Mark-82 (200 pound) or Mark-83 (500 pound) bombs or 2.75-inch rockets and AIM Sidewinder missiles on the recce missions. This was in an era when laser-guided bombs were still in development, so all our bombs were "dumb bombs" in the sense that once released, all they could do was hit the ground somewhere in the vicinity of the target.

The F-4 could carry more than 18,000 pounds of ordnance, but handling the aircraft with a full load was not so much a matter of dealing with the weight as it was the drag. You surely did not want to get involved in air-to-air combat with a full load of ordnance. If you ran into MiGs, you would have to get rid of all the ordnance because you would never be able to take the F-4 into a dogfight with a full load of bombs and rockets. The F-4 was not the most maneuverable aircraft and, at that time, had only missiles for air-to-air combat; the Phantom had no fixed gun for close-in aerial combat. As Norman Polmar and Edward J. Marolda wrote in *Naval Air War: The Rolling Thunder Campaign*, the designers of the Phantom "had sacrificed maneuverability for high-performance and multi-mission capability…The long-range missile's design philosophy that had created the Navy's gunless F-4 degraded the plane's close-in fighting ability." Later models of the F-4 were redesigned to remedy that flaw by including a 20-millimeter cannon.

The standing order for all combat missions was to drop all our unused ordnance before returning to the ship. Under no circumstances were we to bring any live ammunition back to the ship: Not only was it dangerous landing with ordnance still on the aircraft due to a weight restriction, but also there was always the distinct possibility that something might jar loose and explode. Besides, there was no place on the ship to store unused ordnance. "Do whatever you have to do to get rid of those bombs because we've got to clear space for the next shipment," we were told.

On one mission I flew as the wingman for squadron executive officer Rudy Krause. We were ordered to look for the always available "targets of opportunity" in North Vietnam. Bad weather closed in not long after we went feet dry, so we decided to head back to the boat. The problem was that we both still had full bomb racks.

As we made the turn for the boat, Rudy came over the radio and said, "On our way back, let's look for those triple-masted ships. We know they all have a machine gun on board."

So, we went feet wet and soon spotted a couple of those ships just off the coast.

"Dave, cover me! I'm going in!" Rudy called over the radio.

He rolled in and the sailors onboard opened fire. Rudy released his whole payload of bombs and blew the ship to kindling.

"Now it's your turn, Dave. I'll cover you."

I spotted my "target of opportunity," rolled in, and was met with bursts from a heavy machine gun. The gunner missed but I didn't. I released my bombs and blew that sonofabitch out of the water.

We were re-supplied with so much ordnance so often there were times when not all of it could be stored below decks and instead was stacked on the flight deck. It was not unusual to see old ordnance dumped over the side to make room for new stuff coming in; we had so many bombs and rockets we could not use them all. You did not have to be an admiral to figure out somebody in the supply chain did not know what he was doing.

Our recce flights usually consisted of two planes, which is a section. My RIO was Lieutenant (junior grade) L.J. Filbert. He was always a bit nervous going into combat, so every time we launched from the carrier I would yell, "Oh, shit!"

Filbert would come over the radio yelling excitedly, "What's wrong! What's wrong!"

He obviously did not appreciate my strange sense of humor.

On missions I was usually the section leader, meaning I was the lead aircraft. My wingman most of the time was a man I'll refer to only as John to save his family any embarrassment if any of them ever read this. John was a good guy, but he was always scared shitless when we went on missions. If we flew south off the coast, he was off to my left, away from land. If we flew north, he was always to my right, again away from land. I was always between him and the anti-aircraft guns because of his fear of being shot down. I knew that if I ever got into a tight spot, I couldn't depend on John to bail me out; he just wasn't suited for that type of job, and I don't blame him. I eventually told him he needed to get out of combat flying.

When flying those two-plane missions we generally were in a formation known as the "Loose Deuce." Essentially, in this formation if an enemy plane is spotted, the lead plane attempts to maneuver so the enemy plane chases it. That allows the second plane to position itself for the kill. During my time in Vietnam, though, the North Vietnamese MiGs never really challenged Navy pilots; they were usually flying away from us rather than toward us.

One of my classmates from the Academy with another squadron shot down two MiGs, though. I also knew a pilot, Navy Captain Clint Johnson, who got a MiG-17 kill while flying a World War II-era AD Skyraider, a prop plane, later designated as the AH-1 Skyraider. Sometimes referred to as a Spad, the Skyraider can turn you inside out, which is what this guy did. When the MiG-17 came up behind him, the Skyraider pilot turned in tight, the MiG went past him, and he shot the MiG from behind. It was one of two MiG-17 shoot-downs early in the war by Skyraiders, the first coming in June 1965.

Our biggest concerns as pilots in those early years of the war were anti-aircraft artillery (also referred to as Triple A)—which the North Vietnamese had in great abundance, thanks to China and the Soviet Union—and SAMs. I had a few close calls with SAMs but was able to avoid them by out-maneuvering them before they got me. I had the same good luck when it came to the anti-aircraft artillery.

Naval aviators have a tendency to turn left because all our traffic patterns leaving and coming back to the carrier are left turns. We were told before we shipped out that the North Vietnamese were aware of this tendency, so when we pulled off a target we should turn right instead of left. If the NVA were shooting at you with anti-aircraft artillery and trying to figure out which direction you were going to turn, they were likely to think you were going to turn left. So, we would turn right.

With SAMs, the best way to defeat them was to make either a high G descent or a high G turn. If you did that, the missile went right past you because it was going so fast it couldn't track you. Then you'd hear an explosion because the NVA would detonate the SAMs from the ground, hoping that the debris from the explosion would hit your aircraft. From experience, I can tell you that a SAM looks like a high-speed telephone pole coming at you when it's fired.

The Navy attack aircraft often carried SHRIKE missiles on their missions. If those pilots had a SAM coming after them, they would launch a SHRIKE missile, and the SHRIKE would home in the NVA radar station. When the radar operator saw that SHRIKE launch in his direction, he would shut down the radar site, and the SAMs would be useless. This scenario did not always work, and those SAMs were able to shoot down a fair number of our aircraft.

The possibilities of being shot down and captured were always somewhere in the back of the minds of just about every pilot who flew missions over North Vietnam. To prepare for those possibilities, all of us carried a number of individual survival items that ranged from survival radios to flares to signal panels to our individual handguns with a few extra rounds of ammunition. In addition, we carried what were known as a SEEK-2 (Survival, Evasion, Escape Kit). These were two 7 ½ by 5-inch packs that could fit into one of the pockets on the survival vests we wore. Here are just a few of the items that fit into those little packets:

- Liquid soap
- Sun and bug repellent
- Folding razor knife
- Wipe-on towelettes that could be used as toilet paper or to start a fire
- Aspirin
- Diarrhea tablets
- Anti-malaria tablets
- Anti-infection tablets
- Amphetamine tablets
- Salt tablets
- Motion sickness tablets
- Mirror
- Waterproof matches
- Sunglasses
- Combination hacksaw and knife blade
- Water purification tablets
- Wrist compass
- Mosquito head net and mittens
- Flashlight and cord
- Sewing kit
- 1 large Tootsie roll

And that's only about half of what was in those little packs. Thankfully, I never had to open mine.

Our first line period at Yankee Station ran from August 10, 1966, to September 12, 1966. The Tophatters lost neither aircraft nor pilots during that period, but several other squadrons in the air wing were not as fortunate. The wing lost four aircraft although the pilots of three of them were rescued. One pilot, Lieutenant Norman L. Bundy of VFP-62, died when his RF-8G Crusader crashed at sea while he was on his way to North Vietnam on a photo reconnaissance mission. His body has never been recovered.

Although my memory is a bit foggy about when this next event occurred, I believe it was sometime during that first line period. More than 50 years have passed, but I have told very few people about it because of its sensitive nature and what it might have done to an admiral's career had I made this public knowledge. I'm relating it here because it says so much about the manner in which the Vietnam War was conducted and just what the brass who were supposed to be running it—but were not—thought of it.

During that first line period, the *Roosevelt* was the flagship for Rear Admiral Ralph Cousins, the acting commander of CTF-77. One day I was in the *Roosevelt's* Ready Room preparing for a mission when this voice came over the intercom, "Admiral Cousins would like to see Lieutenant West on the flag bridge now."

It was sort of like hearing a voice coming out of the sky saying, "God would like to see Lieutenant West now." As I got up and started walking toward the flag bridge, people stared at me and asked, "What in the hell did you do now?"

I had no idea why an admiral would want to see me although I *had* heard the name Ralph Cousins before. After my second year at the Academy, I went home for a few weeks during the summer and my father asked me, "Dave, do you know a Captain Ralph Cousins?"

"Who, pray tell, is Ralph Cousins?" I responded.

My father went on to tell me that Cousins grew up across the street from him in Ironwood and that they knew each other well.

Cousins had returned to Ironwood that summer for the July 4 celebration and reconnected with my father.

"Dad," I said, "Midshipmen at the Naval Academy don't talk to Navy captains unless talked *to*."

I thought nothing more about the exchange in Ironwood with my father and had only a vague sense when we got to Southeast Asia that Cousins was on board the *Roosevelt*. I was too busy trying to do my job without getting shot out of the sky to worry about who was running the air war or the task force. But now I was concerned I had done something that I did not remember doing that required the personal attention of a two-star admiral. My walk to the flag bridge that day was almost like a walk to the gallows.

When I reported to the bridge, the admiral greeted me courteously and tried to make me feel at ease by talking about Ironwood and my father. He was a trim, distinguished-looking man who presented himself as an admiral should. It was only later I learned he also was a decorated naval aviator, having been awarded the Navy Cross and two Distinguished Flying Crosses for aerial combat in the Pacific during World War II.

Then, he popped a question I certainly was not expecting.

"What do you think of this war, Dave?" he asked.

I was caught off guard by his candor. At first, I thought he was kidding. An admiral never asks a junior officer his opinion of anything, much less about an unpopular war. Junior officers are just supposed to keep their mouths shut and do what they are told. I swallowed hard before I answered.

"Well, Admiral," I responded, "I think it's the best war I've ever been in."

He asked me if we were completing the proper recces, whether we were hitting the right targets, and whether we were using the proper weapons.

"What do you think about what we're doing with our weapons systems?" he asked.

Of course, I could not keep my big mouth shut, so I told him exactly what I thought.

"Well, Admiral, quite frankly I don't understand why we're wasting bombs and rockets on bamboo huts and bamboo bridges. That's a waste of money and a waste of time, and we're going to get people killed."

"Do you want to know why we're doing that?"

"Yes, sir, I would like to know."

Cousins turned and looked toward the east, pointed his finger, and said, "Because that asshole in the White House won't let us run this war the way it should be run."

I swallowed hard and thought, "Why is he telling me this?" I figured this guy's career was going to come to a quick end with talk like that, especially to a lowly lieutenant like me.

Over the years I have tried to figure out why a rear admiral with so much power and so much at stake would confide in me as he did that day. The only thing I have ever been able to figure out was that he considered my father a good friend, and he trusted me — like father, like son — with his most sensitive feelings about how the war was being conducted, even though he knew that if anyone ever found out what he had said, it would likely mean the end of his career.

I am sure I am not the only one Cousins confided in about these things, but I suspect the others were of similar rank and stature as he.

Whatever Cousins' negative feelings about the war might have been, they did not hinder his career. He went on to pick up his third and fourth stars and eventually became Deputy Chief of Naval Operations, the second-most senior admiral in the Navy.

Before retirement, he spent three years as commander of the U.S. Atlantic fleet and was the supreme allied commander of all NATO forces. He also was on a long list of candidates under consideration to be head of the Central Intelligence Agency under President Gerald Ford in 1975.

Cousins died in 2009 at the age of 94 due to complications from a fall, as had my own father in 1970.

I kept that conversation with Cousins to myself for years. But, after that meeting, I became more skeptical than I already was about how the war was being run and what our goal was. I never saw or talked to the admiral again, but that conversation remains as fresh today as it was in 1966.

The *Roosevelt* left the Tonkin Gulf on September 12 and headed for NAS Yokosuka, Japan, for a port call. From there the ship went to NAS Atsugi, Japan, for some repairs before we returned to Yankee Station on October 2. We were back only a couple of days when the ship lost a blade on one of its propellers and we had to return to Japan for more repairs. The Tophatters went to Atsugi to continue flying while the *Roosevelt* was in dry dock.

We finally got back to Yankee Station on October 20 and immediately began combat operations in our second line period. This time the air wing either was not as lucky as it had been during our first line period, or the North Vietnamese had become more proficient with their missiles and anti-aircraft artillery.

On October 20, Lieutenant (junior grade) Frederick R. Purrington of VA-172 was shot down and taken prisoner. On November 1, Lieutenant Allan Carpenter of VA-72 suffered the same fate when his A-4E Skyhawk was shot out from under him while he was returning from a bombing mission. Purrington and Carpenter were held as POWs in North Vietnam until 1973.

Three days after Carpenter's capture, eight members of the

Roosevelt's crew were killed when a fire broke out in a storage area that held oil, hydraulic fluid, and other chemicals four decks below the hangar deck. The eight apparently died of asphyxiation after they took refuge in a compartment next to the burning chemicals.

On November 12, just as we were about to finish our second line period and head to the Philippines for some time off, the air wing lost two more pilots and two more aircraft. VF-12 skipper Commander Robert C. Frosio and his wingman, Lieutenant James G. Jones, were returning from a night mission when they collided during their approach to the ship and crashed at sea. Both were killed, and their bodies were never recovered.

Jones was awarded the Distinguished Flying Cross posthumously for a mission the day before he was killed when his missile strikes took out an anti-aircraft artillery battery during a 40-plane bombing mission on a rail yard south of Hanoi.

When the *Roosevelt* got to Subic Bay on November 15, 1966, the air wing again went to Cubi Point, where we got a little bit of rest and did a lot of relaxing. For many, that meant spending a good deal of time at the O Club trying to forget the war and being so far from home. Our conduct there was not always in keeping with what polite society might consider statesmanlike, but we were amped-up naval aviators looking for a good time and would not be denied.

One of our favorite pastimes was strapping a young pilot to a chair set on rails and then propelling him out the front door and down the front steps. We called it "catapult practice." Most other clubs would have frowned on the practice, but we were naval aviators, and who in their right mind would want to go to a bar with a bunch of drunk naval aviators besides other naval aviators?

One night I walked into the O Club bar at Cubi Point and saw my buddy from the Academy, Denny Moore, standing at the bar, already well-fueled with the 10-cents-a-shot liquor they served. I knew Denny had just been turned down for compassionate leave he'd requested after his wife, Kim, had been injured in an automobile accident.

I figured I would try to cheer him up a bit, so I walked over and said, "How are you doing, Denny?"

He looked at me with eyes that were a bit out of focus and said, "*I'm* fine, but *you're* a cheapskate!"

"What are you talking about?" I asked.

"You didn't buy me a drink," he responded.

I turned to the bartender and said, "Bring him a drink."

The bartender brought over a shot glass and set it down in front of Denny, who looked at it for a minute, and then said, "You know, you really *are* cheap. I wanted two of those."

I told the bartender to give him another shot. Denny drank the first one and then took the shot glass and *Bam!* he bounced it off the thick glass mirror behind the bar.

"What's the matter, Denny?"

"That sonofabitch watered that drink down!" he roared.

Then, he drank the second one and bounced the second shot glass off the mirror. At that point the commander of the air station, a captain, walked into the bar with his wife. He immediately walked over to where Denny and I were standing, and I could tell he was about to give Denny hell.

Before the captain could say anything, Denny turned to him and blurted out, "Why don't you get laid?"

The captain looked as though he was about to respond when Denny turned to the captain's wife and said, "That goes for you, too, lady."

Within minutes, the Shore Patrol showed up and hauled Denny off. Someone somewhere must have intervened because Denny was only restricted to quarters for a few days instead of being thrown into the brig. The captain of the naval station must have heard about the incident as well because we were told he ordered the air wing out of Cubi Point and the *Roosevelt* out of Subic Bay before Admiral Cousins intervened and we were allowed to remain.

On November 18, 1966, two days before we were to depart Subic Bay and return to Yankee Station, VF-14 went through a change of command ceremony. Commander R.C. "Mighty Mike" Adams (of red pencil fame) was replaced by Commander Jack Koach, a 1949 Naval Academy graduate and one of the worst pilots I ever knew. He could not fly worth a crap.

Koach was a real sonofabitch who regarded himself as one of the best aviators on the ship—if not on the planet. He would never miss an opportunity to share stories about his excellent flying skills with the other pilots—although the scuttlebutt aboard ship ran counter to his claims of great flying feats. He was often assigned an LSO to fly with him so that when he came back to the carrier, the LSO could keep him from flying the plane into the water or into the back of the ship.

During one mission over North Vietnam, I dropped my bombs and returned to the ship. In the Ready Room, Koach was there, talking about how he'd dropped his bombs and gotten a direct hit on the intended target. After Koach left the room, the operations officer walked up to me and said, "Good job, Dave."

"What are you talking about?" I asked.

"You were right on target."

"What about the CO?"

"Hell, he was in the next county!"

Whenever an aircraft landed on the carrier, you could hear the boom of it hitting the deck, the screech of the arresting wires, and then the roar of the engines as the pilot applied full power. One day we were sitting in the Ready Room, which is sort of mid-ship, and Koach walked in and announced, "Guys, I'm back."

"Skipper," I said sarcastically, "we *heard* you." Evidently, my dislike of Koach was pretty obvious.

He started to say something, and I said, "Skipper, you don't think I'm trying to brown nose you, do you?"

"Dave," he sneered, "there are a lot of people in this outfit who want to brown nose me, but you ain't one of them."

* * * * *

During those early years of the war, we had to contend with a factor that had very little to do with actually fighting the war. It was the effort to enlist us into trying to sell what we were doing in Southeast Asia to an American public that was becoming increasingly skeptical. We were encouraged, but not ordered, to give interviews to various journalists, including the Hometown News Service, explaining what we were doing and how well the war was going. I wanted nothing to do with that sort of publicity; I did not want my name or hometown mentioned in any news story or news release. When a reporter asked who I was and where I was from, I replied that it was none of his business. I did not want to put my family or myself at risk because protests against the war were starting to ramp up back home, and anyone associated with the military or their families were potential targets for anti-war activists. Instead of promoting myself and my role in the war for public consumption, I avoided all publicity, and I have never been sorry I did.

That included putting myself in for any awards. I was encouraged several times to write up a recommendation for a Distinguished Flying Cross for myself, but I considered that self-promotion, a trait with which I was not comfortable. A number of other pilots did not mind doing that, however, and while some of those DFC's were well-earned by the recipients, a number were not. I always thought nominating yourself for any award defeated the purpose for which it was intended: Awards were for recognition from others for what you did, not what you thought you deserved.

As I have told a number of people over the years, there were two medals I never wanted to be awarded: the Medal of Honor and the Purple Heart. The first brought a certain celebrity status and notoriety that I could do without; the second brought pain.

One recommendation to put myself in for a DFC may have followed a mission in which we sent in a strike package of 16-18 planes to take out a SAM site near Haiphong. Someone up the chain of command wanted to get photos of this site after the strike so the pictures could be released to the media to demonstrate what great work we were doing. About five minutes before the launch, the air group commander came to me and said, "Dave, the photo plane is down, so we're going to give the guy in your back seat a hand-held camera. We want you to go in last and take pictures of the bomb damage assessment."

Normally, we would do our bombing runs at about 5,000 or 6,000 feet, which was a bit low for the SAMs to have a good chance of getting us and a bit high for the anti-aircraft artillery. I don't recall at what altitude I was supposed to make that photo run, but it was much lower than the normal bomb run. Whatever altitude it was, I went over that target at about 600 miles per hour. I did not care one bit whether my back-seater got any usable pictures; I just wanted to get in and out as quickly as I could.

Our third line period started November 24, 1966, Thanksgiving Day that year. Despite repeated pleas from top Navy, Air Force, and Marine Corps brass to loosen the restrictions under which we were flying, they were to no avail. We were still prohibited from attacking any targets within 30 nautical miles of Hanoi, 10 nautical miles of the port city of Haiphong, and 25 nautical miles of the Chinese border. There was a great fear in Washington that if we did too much, we were likely to raise the ire of Communist China, which would then send a few million soldiers screaming across the border as they had in Korea little more than 15 years earlier.

The only thing these flight restrictions accomplished, though, was to give the North Vietnamese time to build up their SAM and anti-aircraft artillery sites; although they did not always know when we were going to fly, they knew where we were going to fly—and more importantly, where we were *not* going to fly. It got dicier and dicier every time we flew a mission.

On December 2, 1966, the combined air wings from the *USS Roosevelt* and the *USS Ticonderoga* teamed up with attack aircraft from the U.S. Air Force to launch a 200-plane strike on a military vehicle depot south of Hanoi. As I recall, there was a North Vietnamese MiG base nearby with all these airplanes parked along the runway with absolutely no protection. But we were told, "Don't destroy the aircraft on the ground." I thought that order was kind of dumb: Here we were wasting a lot of ammunition on bamboo huts and bamboo bridges, but we were not allowed to hit the military assets of North Vietnam. We were told we had to have permission from the Joint Chiefs of Staff (JCS) to go after those enemy aircraft.

It was only much later that I learned the real truth, which was that the air war was being run from Washington by civilian politicians, not the JCS. There was no long-term strategy for the air war. Instead, the so-called strategy was being developed piecemeal week by week and month by month. We were told more often what we were *not* supposed to do rather than what we were *supposed* to do. The chief architects of that strategy were President Lyndon B. Johnson and Secretary of Defense Robert McNamara. Johnson liked to boast, "They can't bomb the smallest shithouse without my approval."

One admiral called this type of warfare "targeting by remote control." It made absolutely no sense for the politicians in Washington to have that much say-so over how to run the war. Those of us doing the fighting could do little more than salute and march on.

On that December 2 mission, CVW-1 lost two more pilots and two more aircraft, including the second commanding officer of VA-172 during the deployment. Commander Bruce A. Nystrom and his wingman, Ensign Paul L. Worrell, were shot down. Worrell's remains eventually were recovered and returned to the U.S. in 1985, but Nystrom has never been accounted for.

Two weeks later, we went back to that same truck depot, this time with about 40 aircraft in the strike package. The air wing lost two more aircraft and four more men. Lieutenant Commander Claude Wilson of VA-72 was shot down and killed; his remains were found and returned to the U.S. in 1988. Three of five crew members of an E-1B Tracer, an airborne early warning radar aircraft from VAW-12, were killed when the pilot was forced to ditch at sea. They were Lieutenant (junior grade) Gerald Holman, Lieutenant (junior grade) Richard Mowrey, and Lieutenant James Murphy. Of the three, only Murphy's body was recovered.

Our third line period officially ended December 27, 1966. I recall very little about those last two weeks of missions; I don't even remember that on December 26, Bob Hope brought an abbreviated version of his famous Christmas show to the *Roosevelt.*

Hope had performed earlier in the day on the carrier *USS Bennington*, which was steaming nearby. Hope and some of the cast of the show came to the *Roosevelt* by helicopter and put on a performance on the hangar deck for the crew. Among those who performed with Hope were singer Vic Damone, comedian Phyllis Diller, actress and dancer Joey Heatherton, and singer Anita Bryant. I must have been in my stateroom asleep or playing cards.

We finally departed Yankee Station on December 28 and headed for Subic Bay, The Philippines. The ship was scheduled to go from there to Hong Kong before returning to Subic Bay and then anchor in Cape Town, South Africa, for some R&R on its way home to Mayport.

I managed to avoid that long voyage home, however. The military had chartered a civilian aircraft for personnel returning to the U.S. from Southeast Asia and had a limited number of seats available for the Navy. All pilots and RIOs were told to write their names on a slip of paper and put it into a hat. Those whose names were pulled in the lottery would be able to fly back to the States in civilian comfort instead of having to take that long sea voyage. It just so happened my name was the first one pulled out of the hat, and I gratefully took the opportunity to get home more quickly. The air wing commander sat behind me on the flight home, and I think he drank the whole way.

On that flight, I spent some time reflecting on what we had done. VF-14 did not lose a pilot and lost only one aircraft, and that was on a training flight in Japan following our second line period. In that incident, the engines of an F-4B flamed out and the pilot and his RIO ejected. Both were rescued, and the pilot had only minor injuries while his RIO suffered a serious back injury.

As a squadron, we were good, dumb, or lucky during our tour in

Southeast Asia, but, as a pilot, I would rather be lucky than good any day. For me, it probably was just a little bit of all three. Even in that environment where I knew I could be shot down at any time or the aircraft could malfunction and I would have to bail out, I thoroughly enjoyed the flying. It did not even bother me too much when I was being shot at—as long as I did not get hit.

According to squadron records, during the roughly six months we were in Southeast Asia, VF-14 flew 1,137 sorties and had 830 day landings and 275 night landings on the carrier. I'm not sure why those numbers do not add up, but those are the numbers that were submitted to the CNO by Jack Koach. We also dropped 863 bombs and fired 9,865 rockets as part of the Rolling Thunder campaign although I am not sure how much damage we did.

What did we accomplish? I'm not sure we accomplished much of anything except put numbers on a board for the politicians in Washington to manipulate for their own purposes. I was just glad to put Vietnam behind me and hoped I would never see it again. While many naval aviators often made multiple tours in Southeast Asia, the Navy, thankfully, had other plans for me.

CHAPTER 6
JUMPING SHIP

The Miramar Nightmare and My
Return to Civilian Life

By the time I left Southeast Asia and arrived back in the United States in early January 1967, anti-war sentiment was creating a distinct fault line not only in America, but throughout much of the world. Anti-war protests were rapidly gaining in frequency and crowd size, as was the level of animosity directed at those of us in uniform. Hostility was being directed against the warriors and not just the war.

It was a particularly trying time for those of us committed to serving our country. While we had taken an oath to preserve and defend the Constitution, some, including me, were beginning to have doubts about the war's larger purpose. We also were wondering whether anyone in Washington knew what they were doing, as Admiral Cousins had so bluntly put it to me months earlier on the flag bridge of the *USS Roosevelt*.

That conversation with Cousins continued to rattle around in my head. I still could not figure out why he had been so forthcoming to a junior lieutenant. He may have known my father, but he did not really know me. I still was not sure if he was testing me or just letting off steam. It had been a strange, almost surreal, conversation. His candor, though, only reinforced my own doubts about just what we were doing in Vietnam.

When I got back to the States, I was reassigned from VF-14 to VF-121 at NAS Miramar just north of downtown San Diego. VF-121, known as the Pacemakers, was not the typical carrier squadron comprised of 12 planes and 12-14 pilots; it was a Replacement Air Group (RAG), the West Coast equivalent of VF-101, where I had learned to fly the F-4B Phantom in Key West. This RAG had 70-80 airplanes at any given time and somewhere in the neighborhood of 400-500 pilots and sailors. It was a large outfit with a lot of people and numerous moving parts, all of which required constant attention.

I had two duties with the Pacemakers, both considered primary duties: One was to train incoming pilots to fly the F-4. The other was as maintenance officer in charge of four of the systems on each of those 70-plus aircraft in our inventory: propulsion, seats, airframes, and structures. Each of those systems also had a junior officer overseeing it who reported to me. I, in turn, reported to my superior.

VF-121 also was a cannibalization outfit. If a squadron was going to sea and had an airplane that we referred to as a "basket case" because it had many missing parts, that airplane was flown down to us and dumped on the tarmac. Then, the same squadron that dumped the plane on us took one of our fully operational airplanes because it made no sense sending an airplane to sea that was not flyable. So, we had one fewer good plane and one more basket case we had to fix so it could fly again.

The operational tempo of the Pacemakers was like a high-speed nightmare that lasted all day and into the night. Since I was a maintenance officer, I was expected to be on hand for quarters (the daily muster and inspection) at 7 a.m. I would perform my maintenance duties and often was not scheduled to fly until late afternoon. That meant after my flight I had to debrief and often ended up not getting home until after 10 p.m. The pace was so frantic it was almost like being deployed—except there was even less rest than when we were flying combat missions.

In 1967 and 1968, the Navy's demand for pilots was so great we were flying seven days a week to get crew members ready to ship to Vietnam. One of the pilots I instructed during that time was Navy Lieutenant David M. Walker, a native of Columbus, Georgia, and a 1966 graduate of the Naval Academy. He eventually went on to become a NASA astronaut and flew on four Space Shuttle missions over two decades, starting in 1984. Walker was not the worst pilot I ever trained, but he certainly was not one of the best. One day we flew out to Point Mugu, California, to shoot some missiles. We were in a four-plane formation, and he damn near ran over me. I don't think he ever saw me.

More than 20 years later, he did the same thing while flying a T-38 Talon to Washington, D.C., to be honored for his 1989 space flight. He came within 100 feet of striking a Pan American passenger jet on his approach to D.C., and was grounded for three months, costing him command of another Space Shuttle flight.

A number of good pilots went through training with VF-121, though. Two other students of mine were selected to fly with the Blue Angels, the Navy's precision flying unit.

Not all our training was done at Miramar. Once a month we flew to Marine Corps Air Station Yuma, Arizona, for weapons training. That often took a week or more, which meant even more time away from home. That drill became tiresome very quickly.

If you're in a sea-going outfit, you know when you go to sea you will be gone six to eight months and then come home. But with VF-121, it was just like being on a cruise because I was gone all the time. We were not at sea, but we were never home, either going out to the carrier for qualification or going to Yuma for weapons training or tending to maintenance work — the last being a never-ending task.

I sometimes joked with the other pilots that I had to ask my wife what my kids' names were because I could never remember their names since I was never home. That's when I decided I had had enough. I had intended to make the Navy a career, but the pace of operations and the pressure we were under in 1967 and 1968 at Miramar convinced me to start looking for a way out.

I said to myself, "I'm never going to make it in this organization because I don't want to be away from home so much. So, I'm going to get out of aviation. I'm going to get out of it and go someplace where people are sane."

Besides, the handwriting was on the wall: I knew that when my three years of shore duty were up, I would be reassigned to a West Coast squadron that was going back to Vietnam, and I had no desire to repeat that experience.

In late summer 1967, I put in my papers to leave the Navy and become a civilian. I had put in the 42 months I was required to serve after getting my wings and was ready to move on to something else. However, the Navy decided to extend my contract for a year under a policy that specifically applied to Naval Academy graduates. So, I was stuck at Miramar for another year.

I had previously expressed a desire to go to the Navy's test pilot school at NAS Patuxent River in Maryland but had given up on that idea by the time I put in my papers.

Not long after I made the final decision to leave the Navy, a former commander of VF-14, Dick Adams, came to San Diego and I invited him to the house for dinner.

During dinner he looked at me and said, "Dave, I understand you're putting in your papers to get out of the Navy."

"Yes, sir," I replied, "that's what I'm doing."

"I also understand you want to go to test pilot school."

"Yes, sir, but it appears that I'm not going to make it."

"Why don't you pull your papers, and I'll get you in the next class at Patuxent?"

I thought about it for a moment but shook my head and finally said, "No, it's too late for that. Too late."

Although the Navy needed pilots, I was not given any inducement to stay in. Some years later, when my son David was about the same rank as I was when I got out, the Navy was offering pilots $25,000 to resign because the Navy thought there were too many pilots. About 18 months later, the Navy was offering the same pilots $25,000 to re-enlist because it was thought there were not enough pilots. That's the federal government for you!

When I finally left the Navy in September 1968, I did not realize that naval aviation, NAS Miramar, and VF-121 were about to undergo seismic changes and become instrumental in the way American fighter pilots were trained for future aerial combat. Those changes came about because of a study with the rather innocuous title of "The Report of the Air-to-Air Missile System Capability Review." That study later became known simply as the "Ault Report," after its primary author, Captain Frank Ault, a naval aviator and veteran of World War II and Vietnam.

Navy brass commissioned the study after they became concerned about the poor kill ratio our pilots were compiling in aerial combat in Vietnam. Instead of the 14:1 ratio from World War II or the 12:1 from Korea, the Vietnam kill rate was a startlingly low 2.5:1 from 1965-68.

Ault wrote that the missiles on which so many of the aircraft relied in aerial combat, especially on the Phantom, were not designed for traditional aircraft dog fighting. "A primary reason for less-than-desired combat performance of air-to-air missile systems in Southeast Asia is their design optimization for a high-altitude engagement against a non-maneuvering, large (bomber) target. Consequently, they exhibit important limitations in a low-altitude fighter-to-fighter engagement," the report stated.

That—plus the Phantom had no gun for close-in combat.

The bottom line was that the Navy needed more reliable missiles and better close-quarters combat training for its pilots. As a result of that report, in March 1969 the TOPGUN School at Miramar was created. VF-121 was the host squadron for the Navy's best pilots and their aerial combat training, later immortalized in the 1986 Tom Cruise movie *Top Gun*.

By the time the TOPGUN school was up and running, I was well into my first civilian job after leaving the Navy. I had interviewed with four or five different companies while waiting for my release from active duty. I finally went to work for J.T. Baker, a New Jersey-based chemical company, as its sales representative in Wisconsin, the Dakotas, northern Michigan, and northern Illinois.

I had no real regrets about leaving the Navy: I had served my time faithfully and to the best of my ability. I enjoyed the time I spent flying, even in Vietnam. I wanted more challenges, though. I briefly considered applying to become an astronaut, but dismissed the idea because I did not exactly fit the ideal model of an astronaut. Even back then, I was not especially politically correct and spoke my mind when I saw something that I thought was not right.

In my first year with J.T. Baker, I was able to increase sales in the region by 10 per cent. I was good at what I did, but I could not get my mind off flying. And, as a member of the U.S. Navy Reserve, I was able to do it as often as I liked. As much time as I devoted to my civilian job, I spent just as much time at NAS Glenview in suburban Chicago flying the A-4C.

I finally got to the point that I had to be honest with myself and my family about what I really wanted to do with my life. I said to myself, "This is not going to work. I'm going to have to get back into aviation or I've gotta get the hell out of aviation. I can't do both."

While I was trying to figure out what to do and how to do it, one of my wife's relatives who had an interest in aviation called me one day. (I was never really sure what interested him about aviation because he sold pressure cookers for a living. Maybe he was living vicariously through me and just wanted me to get back in the air.)

"Dave," he said, "are you aware that *Aviation Week* has an ad from the FAA that says highly experienced, high-performance jet-time pilots with a minimum of 1,500 hours are being sought to send to test pilot school in Mojave, California? I thought you might be interested."

I was more than interested. It was the tipping point for which I had been searching.

"Give me the address for that," I asked eagerly. The next day I sent in my résumé, and it was only a few days later that I received a call from a representative of the FAA asking if I was still interested in attending the test pilot school.

"Yes, absolutely," I responded.

"We've got you enrolled in the next class," she said.

I gave my notice at J.T. Baker and left the company in January 1970. That same month I drove to California and reported to the Air Force Test Pilot School at Edwards Air Force Base in the high desert community of Mojave, California. I was not quite sure where this would all lead, but knew if I did well at the school, I would be doing not only what I wanted to do, but what I felt I had been born to do. With great anticipation, I took the next step in my career in aviation.

CHAPTER 7
TEST PILOT SCHOOL AND
THE FAA, ROUND 1

When I informed my boss at J.T. Baker in January 1970 that I was leaving the company to take a job with the Federal Aviation Administration and return to flying, he wrote a letter to me expressing his appreciation for what I had done and wished me well, even though I had worked there little more than a year. "We had big things planned for you and we intended to promote you very quickly," he wrote. "We're sorry to see you go."

As good as his comments made me feel, there was no looking back or reconsidering. Once I made up my mind to attend test pilot school as an FAA employee, I don't think anyone could have said anything to make me change my mind: I wanted to be a test pilot and that was the end of any argument or discussion. As with so many other things in my life, once I decided to do something, whether it was recovering from that serious hand injury as a youngster, graduating from the Naval Academy, or becoming a naval aviator, I was going to reach that goal. If that is what people refer to as being goal-oriented, then I was goal-oriented, especially when it came to flying.

The decision to leave J.T. Baker meant I had to uproot my family once again and move from our home in Illinois to California. Before leaving the military, we had lived in Jacksonville, Florida; Key West, Florida; Norfolk, Virginia; and San Diego, California. Frequent moves are normal for a military family, though, and since my sons David and Travis were only 4 and 2, respectively, at the time, picking up and moving to California with kids that young was no big deal.

Our new home was in the high, dry, desert city of Mojave, about 90 miles north of Los Angeles and just a few miles from the gates of Edwards Air Force Base, home of the U.S. Air Force Test Pilot School. I was one of only two civilians at the school; the rest were Air Force pilots, some of whom eventually were selected by NASA for the space program. The civilian portion of the school lasted nine months while Air Force pilots stayed an additional three months.

At test pilot school you learn how to evaluate an airplane's handling qualities and performance. The curriculum is designed to teach students how to test airplanes for certification either for military or civilian use, the latter the responsibility of the FAA. Technically, the two types of aircraft are totally different in what they are designed to do, but obviously there also are a lot of similarities. Most of the planes we flew at the school were military aircraft because the big civilian aircraft manufacturers like Boeing, Beech, and Cessna had their own test pilots.

Our instruction days usually were split—half classroom work and half flying. We generally flew in the mornings when the high desert air was clear and the winds relatively calm. The afternoons were reserved for classroom work with courses as tough, if not tougher, than anything I had encountered at the Naval Academy. There were classes in all the specifications of each of the aircraft we were going to fly in addition to work in thermodynamics and advanced aerodynamics. It was like studying for an advanced degree in aeronautical engineering.

One of the instructors was an Air Force major who also held a PhD in aeronautical engineering from the University of Alabama. He taught a course he called "Walking Around the Pasture." It involved taking an electrical diagram of an aircraft and evaluating the system simply from the diagram.

I must have had a funny look on my face one day during one of his lectures because he stopped what he was doing, looked at me, and said, "Dave, did you have such-and-such course in college?"

I said, "Doc, I'm not sure, but when I get home tonight, I'm going to check my sheepskin to see if I had it."

He smiled and said, "If you have to check your sheepskin to see if you had it, it's not going to help you."

We flew a variety of aircraft at the school to acquaint us with both single-engine and multiple-engine models. I had one flight in a Boeing B-52 bomber and one flight in a North American F-100 Super Sabre. I had 30-40 hours in a Lockheed F-104 Starfighter, plus a number of hours in the Martin B-57 Canberra and the Northrop T-38 Talon. In fact, most of my performance time was in the T-38, which also was used primarily by those Air Force pilots headed to astronaut training.

I even had a couple of flights in a Convair F-106 Delta Dart, the Air Force's primary all-weather interceptor at the time. That was sort of an "Aw, gee whiz" treat for me. Even so, I still had to study the emergency procedures for ditching the aircraft and getting out if something went wrong. The emergency procedures for the F-106 were about as long as *War and Peace*—just recovering from a spin involved completing 15 or 20 different steps.

I went to my instructor before my flight and said, "Why in the hell is this list so long?"

"Well, Dave," he said, "it's designed to take your mind off the aircraft while you're going through the checklist so that by the time you're finished reading, the aircraft will have recovered itself." The F-106 had good handling qualities, as I recall, but it had that big delta wing that did not make it a particularly attractive aircraft.

Learning to fly an airplane is one thing; testing an aircraft to determine its performance capabilities is quite another. At test pilot school we were learning how to get performance data from an airplane and how to evaluate its handling qualities. An airplane is not much use to a pilot if that pilot does not know its characteristics, such as its stall speed, its minimum control speed, how high and fast it can fly, its rate of climb, and any little idiosyncrasies it might have.

For example, to determine the stall speed of an aircraft, we did a series of stalls in different configurations from clean to dirty—"clean" meaning the landing gear and flaps are up, and "dirty" referring to landing gear down with a variety of flap positions. These speeds are referred to as V-speeds, which are determined by aircraft designers, engineers, and manufacturers during type-certification testing.

There are dozens of V-speeds for each type of aircraft, and each V-speed has a specific number or letter designation. V_1, for example, is the speed at which takeoff can no longer be safely aborted. V_2 is the safe speed for takeoff in multi-engine aircraft with one engine out. V_{MC} is the minimum control speed with one engine out. V_S is the stall speed, or the speed below which the aircraft will pitch nose down, the general indication of a stall. If an airplane stalls, all a pilot has to do to recover is apply power.

With civilian aircraft, one of the key V-speeds is V_R, the rotation speed, or the speed at which the pilot is able to apply power to lift the nose and become airborne. That is not an issue with most military aircraft I've flown. They had afterburners, and with those there is no danger of not getting airborne because you had a lot of thrust. The F-4s I flew had 34,000 pounds of thrust; that's a hell of a lot of horsepower.

But, you still have to determine safe speeds for takeoff and how much runway you need to land and how much runway you need to take off. Any commercial pilot can tell you what the V-speeds are for whatever aircraft he or she is flying because they're in the flight manual, and those numbers are what the test pilot determines through testing the aircraft.

We were learning what we had to do to get to those numbers so that when we tested new or modified aircraft down the road, we would be able to accurately come up with the V-speeds for that particular type of aircraft. We were not trying to match the numbers someone else had already obtained; rather, we were learning how to find a particular number for a particular type of aircraft. Then, based on the number we obtained as students, we could look at the flight manual and determine whether the number we obtained was a valid number.

You do not take the number in the flight manual and say, "That's what I can do." You're not trying to prove the flight manual is correct; instead, you're trying to prove that you know how to get that number without crashing the airplane. If I did not get the number listed in the flight manual, the instructor probably would have said, "Dave, you're in the wrong school if you don't get the numbers in the book." I got the numbers and graduated from the school in September 1970.

My first posting in the FAA was to the Los Angeles office, located in Cerritos, just south of downtown Los Angeles. Since I was the junior pilot on the block, I got most of the cats and dogs, meaning I got most of the aircraft tests nobody else wanted to do.

At that time, the L.A. office was testing the McDonnell Douglas DC-10 and the Lockheed L-1011 Tristar passenger jets, but there was never any chance I would be part of those tests. They would never put someone as inexperienced as I was on major programs like those because I would not have known what I was doing. Instead, I had to work my way up, essentially starting from the bottom of the aircraft barrel. However demeaning it might have seemed, it was a good learning experience because I had to finely tune my aviation instincts to avoid trouble.

One of the first projects I had was in Tucson, Arizona, where Hamilton Aviation was buying up old twin-engine Beeches, taking the reciprocating engines out of them, and putting on turboprops. I was sent there to test the aircraft to see if they would meet FAA certification following the engine changes.

When you conduct flight tests, you're required to do them with the center of gravity at certain positions. One test involves the center of gravity the most aft you can fly the plane, and one involves the center of gravity the most forward. On my first day there I went to look at the airplane and when I opened the door and looked in, I saw what appeared to be about 300 pounds of sand in the rear of the plane.

I got in touch with the company pilot because I knew there were going to be some problems if that sand was not moved.

"You better get that weight and balance guy here because somebody screwed up," I said.

"What's the problem?" he asked.

"This airplane isn't going to fly with 300 pounds of sand back there."

The company pilot said the flight test engineer Hamilton had hired for the program loaded the sand in the aircraft. I told the pilot to bring the engineer out to the plane. When I explained the problem to the flight test engineer, he started going back over his numbers and discovered his calculations were wrong. The center of gravity for that plane with the sand loaded was about 10 inches aft of the most aft allowed, which made the center of gravity originally computed by the engineer to be outside the allowable limit for that aircraft. We never would have gotten off the ground with all that sand so far aft, so the sand was moved forward.

What this experience taught me was you had to be careful of some of those flight test engineers because all they were looking for was their $30 an hour.

Working for the FAA and testing civilian aircraft rather than flying military aircraft as I had done in the Navy required a bit of a different mindset. Civilian aircraft are designed for performance and, to a certain degree, ease of operation and comfort. Military aircraft, for the most part, are designed for a specific task.

I enjoyed flying military aircraft because they are more mobile, and you can do more things in them without getting yourself into trouble as easily as you can in a civilian aircraft. Test pilots don't perform certain tests for civilian twin-engine aircraft, such as how to get out of a spin because civilian pilots are taught to stay away from those sorts of things. I enjoyed flying civilian aircraft, such as the DC-9, which is a nice-flying airplane. But it is much more fun flying an F-5 than it is a DC-9: it's the difference between driving a sports car and driving the family station wagon.

Still, once you get into the business of testing airplanes, whether military or civilian, among the issues you are evaluating are its handling qualities. Is it responsive? Does it go where you want it to go? Does it do every function with ease, or do you have to fight the controls all the time? Every time you fly a different airplane, you've got some things to learn that you did not know before you flew that particular plane so that you can honestly and accurately draw conclusions about its handling qualities.

One of the other major differences between flying and testing military and civilian aircraft is that the FAA uses FARs, (Federal Aviation Regulations) as its guidebook on everything related to aviation. It has thousands upon thousands upon thousands of details governing everything from fixed-wing and rotary-wing aircraft design and maintenance to monitoring model rocket launches to ballooning to pilot training to model aircraft operations.

The military, meanwhile, generally does it somewhat more simply. If the military wants an aircraft for a particular mission, it generally puts out a Request For Proposal, listing in detail what it wants that aircraft to be able to do. Once that RFP goes out, the aircraft manufacturers must decide whether they want to spend the money to design an aircraft that meets those requirements. If they decide to take the risk to design that plane, they also are risking the possibility that their design will not be selected by one of the military services, as happened a few years later when the Air Force was in the market for a close air support aircraft, a project in which I later became involved.

An exception to that was the Northrop T-38, the first supersonic trainer used by the Air Force as its primary jet trainer. In the early 1950s, Northrop began developing—on its own dime—a small, supersonic jet that could be flown off Navy aircraft carriers. When the Navy decided not to buy into that project, Northrop continued spending its own money to develop the fighter for foreign markets.

When the Air Force went looking for a new training jet in the mid-1950s, Northrop again shifted directions and out of it came the T-38. Not long after that, the T-38 served as the template for Northrop's F-5 Tiger fighter jet. All the models of the F-5 have good handling qualities, and it is easy to see the similarities in the airframes of the two aircraft. Like the T-38, the development of the F-5 was privately funded by Northrop.

My first year as a test pilot for the FAA was almost my last year as any kind of pilot—and of me. It happened during a cross-country flight from Los Angeles to Rhode Island to see my mother in the summer of 1971.

I was flying an A-4 Skyhawk with extended-range tanks when I made a stop at NAS Glenview (Illinois) for fuel. After fueling up and figuring how much the plane weighed, I calculated my go/no-go point, which is based on whether you are able to reach a certain speed with 3,000 feet of runway remaining. If you are not able to get that speed, you have to abort the takeoff.

I based my computations on Glenview's north-south runway, which was about 9,000 feet, compared to the east-west runway which was about 6,000 feet; beyond the runway were some housing developments and Lake Michigan. As I was getting ready to roll out, the tower changed the duty runway from north-south to east-west, but I did not re-compute for the shorter runway. It was also a hot and muggy day, and that airplane did not like to fly in hot weather.

I started out and was rolling, rolling, rolling down the runway with a full load of fuel, and I was nearing the end of the runway, and I said to myself, "If I don't get airborne now, I'm *not* going to get airborne!" So, I yanked back hard on that stick, and I don't think I cleared those houses by more than 10 feet. Later, I castigated myself, "What a dumb thing to do!" It was not my greatest moment in flying; I was lucky to get that plane off the ground. And, as I have said many times before, I would rather be lucky than good any day of the week.

My luck did not extend to job security with the FAA, however.

As I was approaching the one-year anniversary with the agency, I began to hear disconcerting news from the aircraft industry. There was a downturn in business in 1971. Aircraft manufacturers were laying off people and shutting down projects. That summer my boss came to me and said the FAA was going to have a RIF, or reduction in force. Although I was among the most junior pilots in that office, they would keep me on because of the agency's veterans' preference program. But, I was informed they would likely have to make me a flight test engineer.

"No, you're not," I told my boss. "I did not go to test pilot school to be a flight test engineer."

"What do you propose to do?" he asked.

I said I was going to write to the FAA and request they release me from the four-year contract I had signed in exchange for the FAA paying for me to go to test pilot school. If they were not going to allow me to be a test pilot, I figured they were not keeping their end of the bargain. My boss did not think the FAA would buy that argument but said I should write the letter and see what happened. So, I wrote to the FAA, making the argument that our original agreement had been for the agency to send me to test pilot school to be a test pilot, not a flight test engineer. If the FAA was not going to stand by the contract, it should release me from the contract. The FAA admitted my chances of staying on as a test pilot were close to zero and gave me my walking papers.

Not long after leaving the FAA, I heard about a job with the Hughes Aircraft Company to certify the missile program on the F-14 at Point Mugu, California. They wanted a pilot with a lot of jet experience to fly that program. I went over and talked to them but wasn't particularly impressed with the company's chief pilot and did not pursue the position.

About the same time, I learned the Northrop Corporation (later to become Northrop Grumman) was looking for pilots with extensive time in F-4s. I had close to 2,000 hours in that aircraft, so I gave them a call. I was told their chief pilot, Hank Chouteau, was on a business trip to all the countries whose militaries were using Northrop's F-5 Freedom Fighter and he would call me when he got back. (A number of countries used that plane, so it was a long trip for him.)

Nevertheless, Chouteau called me when he returned and I went to the company's headquarters in Lancaster, California, not far from Edwards Air Force Base, for an interview. Hank and I hit it off well and he offered me a job as a test pilot with Northrop. At that time, the company was developing the YF-17, which would become the template for the F/A-18 Hornet, which is still in use by the Navy. Northrop was also trying to develop a CAS aircraft for the Air Force with the YA-9, which was in competition with the Fairchild Republic YA-10, the latter which went on to become the quintessential A-10 Thunderbolt, better known as the Warthog. (The "Y" designation signifies an experimental aircraft.)

When I told my wife I was taking the Northrop job and we were moving back to the high desert, she was not particularly happy. She wanted me to take the F-14 missile job with Grumman so we could move to Oxnard, California, and live on the beach.

"Look," I reasoned with her, "if I'm going to do what I want to do, I have to go someplace where they have some flying that sounds like fun and that's with Northrop. I have to take this job with Northrop."

Instead of moving to the sand at the beach, we moved to the sand in the desert in Lancaster, California, where the flying *really* was a lot of fun.

IMAGES

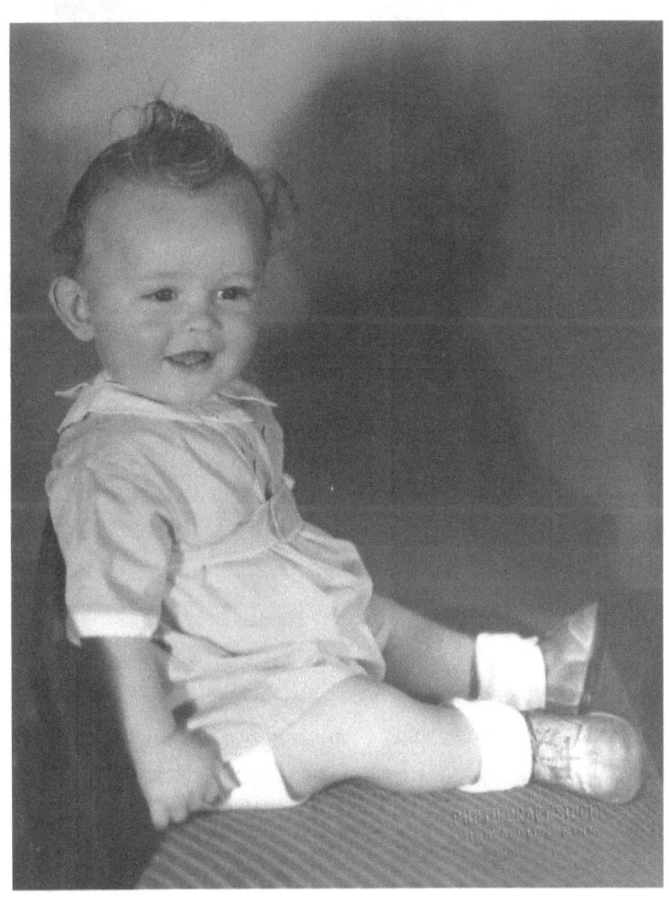

Ready for my close-up at the age of 1 in 1939.

The only known photo of me (age 7 in 1945) with my right hand in a cast following the accident on my grandparents' farm. I endured 21 operations on that hand over a period of 15 years.

I served as class vice president three of my four years at L. L. Wright High School in Ironwood, Michigan. (I'm on the back row, left). This photo is from my freshman year, 1952-53.

I took this photo of my maternal grandmother, Susannah Heczko, in 1957 at the farm she and my grandfather owned outside Ironwood and where I spent a good part of each summer as I was growing up.

At the plotting table aboard the aircraft carrier U.S.S. Intrepid during my summer cruise in 1958 between my Plebe year and Third Class year at the U. S. Naval Academy. It was one of the few times on the cruise when I was able to wear my Navy whites.

Graduation Day at the U. S. Naval Academy. My mother is pinning on my ensign boards while a seemingly disinterested Tecumseh looks on.

Preparing for flight training in a T-2 Buckeye trainer at Meridian, Mississippi.

Receiving my gold aviator's wings in a somber ceremony on December 7, 1962, the 21st anniversary of the Japanese attack on Pearl Harbor.

Serving as a flight instructor in March 1963 at Naval Air Station Pensacola.

*A quiet moment aboard the U.S.S. Roosevelt (CVA–42) at
Yankee Station off the coast of North Vietnam.*

Although I would not talk to reporters or allow myself to be photographed out of concern for my family back in the States, one Navy photographer took this picture of me as I climbed into the cockpit of an F-4. The caption on the back of the photo simply says, "VF-14 pilot mans aircraft for air strike against North Viet Nam."

Visiting with my sisters Ruth (left) and Margaret in Illinois in the summer of 1969.

On the flight line at Northrop in 1973 after another test flight in the F-5E.

The infamous "Navel Quartet." I'm second from the left. It wasn't classy, but it's the type of thing aviators do when not properly supervised.

Members of the Flight Test Branch in the Atlanta Aircraft Certification Office in the late 1980s. From left: Edgar Wilson, pilot; Jerry Boutwell, engineer; Lynn High, pilot; myself; Tom Peters, engineer; Bob Sample, pilot; and Al Sodergren, pilot.

*Rich Adler (right) and me with our Chinese interpreter
during our disastrous trip to China in 1992.*

The Chinese Y-12 – the cause of our disastrous trip to China in 1992.

This photograph accompanied an article in the Bloomberg News about the HondaJet. I'm standing next to the prototype of the aircraft at the Oshkosh (WI) Air Show in 2005. (Jaron Berman photo)

CHAPTER 8
THE NORTHROP YEARS

The four years I spent flying for the Northrop Corporation were among the most interesting, most challenging, and most enjoyable of the 47 years I spent in aviation. There was little of the bureaucratic in-fighting I later encountered with the FAA. I got along well with my boss, Hank Choteau, a veteran of aerial combat in World War II and Korea and recipient of the Distinguished Flying Cross for bravery in combat.

More importantly, the systems and aircraft I helped develop while I was with Northrop were at the forefront of military aviation at that particular time. The job was exactly what I expected the job of a test pilot to be.

My first project with Northrop was the Joint Services In-Flight Data Acquisition Transmission System, or JIFDATS. The purpose of this program was to develop a compatible system for aircraft from all four services that would allow tactical reconnaissance and surveillance information to be transmitted to a ground station in near-real time for analysis—far more quickly than the systems then in use.

Although the Navy was overseeing the program, the aircraft we used initially were an Air Force RF-4C Phantom II and an Army OV-1 Mohawk. The RF-4C was an unarmed intelligence- gathering aircraft similar to the F-4B Phantom I had flown in Vietnam. The Mohawk was a lightly armed twin turboprop primarily used for reconnaissance.

JIFDATS was a complicated project because it involved trying to figure out how to solve numerous conflicts with the transmission of data on different bandwidths from more than a dozen reconnaissance sensors on what eventually would be 10 different aircraft. It was not my job to figure all that out; I just flew the planes with the sensors while someone else looked at the data and tried to resolve the numerous problems.

After about nine months working JIFDATS, I moved to one of Northrop's top projects, the YA-9. By the time I joined the company in August 1971, the program had been in the works for several years with Hank Choteau as the chief pilot. I became the backup pilot on the YA-9.

The YA-9 was Northrop's response to an Air Force request to develop an airplane that could be used for close air support missions. The Air Force encountered a great deal of criticism during the Vietnam War for not having a suitable aircraft to provide CAS for troops on the ground engaged in close combat with the enemy. Most of its aircraft were "fast movers," designed primarily for aerial combat and high-altitude bombing missions. Large numbers of those aircraft were shot down during the war by everything from SAMs to anti-aircraft artillery to small arms fire.

What the Air Force wanted was an aircraft that was a bit slower, more heavily armed, and better able to withstand ground fire, especially from small arms. The closest thing it had during Vietnam that met those criteria was the Douglas A-1 Skyraider, and that was something of a relic from World War II with its single piston-driven engine.

In March 1967, the Air Force sent out a request for information from 21 defense contractors for what it called the A-X aircraft. The general specifications required that this airplane had to be able to maneuver at slow speeds, be able to linger over the battlefield for an extended time, be able to withstand a significant amount of ground fire and be equipped with a rapid-firing cannon. While the initial request specified turboprop engines, that requirement was changed two years later to turbofan engines. The turbofan engines eliminated the propellers and provided more thrust than the turboprop engines.

In May 1970, the Air Force sent out a much more detailed Request for Proposal that specified just what this aircraft should be able to do, including a cost ceiling for each plane. And the bottom line was that the plane was to be built around a 30-millimeter cannon capable of firing up to 4,000 armor-piercing depleted uranium rounds per minute, even though that particular cannon was still in development. Other details the Air Force wanted built into the aircraft:

- Maximum speed of 460 miles per hour.
- Short takeoff and landing (STOL) capability.
- Ability to carry up to 16,000 pounds of weapons externally.
- Cost not to exceed $1.4 million per aircraft.

Six companies submitted proposals, but only Northrop and Fairchild Republic were chosen by the Air Force in December 1970 to build prototypes. While the Northrop design became the YA-9, the Fairchild Republic design was designated the YA-10.

The YA-9 had its first flight in May 1972. I ended up with about 100 hours of test time in it, including weapons testing. The

58-foot wingspan of the YA-9 had 10 hard points capable of carrying everything from bombs to AGM-65 (air-to-ground missiles), but the centerpiece of the aircraft was that cannon. The YA-9 was fitted with a 20-millimeter cannon for test purposes during the development of the 30-millimeter cannon.

While the YA-9 was a conventional-looking aircraft with two underwing engines, the YA-10 was anything but conventional-looking. Its two turbofan engines were mounted at the rear of the plane between the twin tails, making it much less susceptible to ground fire. The cockpit and most of the instrument panel were protected by 1,200 pounds of titanium armor, dubbed "the bathtub" by pilots. And, while its wingspan was only six inches shorter than the YA-9's, the whole aircraft just looked shorter and stubbier, thus eventually earning it the nickname "Warthog."

After a two-month flyoff in October and December 1972, the Air Force chose the YA-10 over the YA-9, and the winner became the A-10 Thunderbolt. It went on to serve with distinction in Operation Urgent Fury (the invasion of Grenada), Operation Desert Storm (the first Gulf War) in Kuwait and Iraq, Operation Enduring Freedom (Afghanistan), and Operation Iraqi Freedom (Iraq again), and subsequent operations in Afghanistan and Iraq.

American Marines and soldiers on the ground in those conflicts came to love the sight and sound of the A-10 when it appeared over the battlefield. The distinctive prolonged ripping sound of the 30-millimeter cannon firing—sometimes referred to as "Hawg farts"—often elicits spontaneous cheers from ground troops because they know that whoever or whatever is at the wrong end of the cannon burst is no longer combat-effective.

Although I never flew the A-10, I thought the A-9 was a great airplane and probably would have performed as well as its competitor. But the Air Force was in a hurry to get the plane into production, and the A-10 was selected in January 1973.

The Air Force also seemed to be in a hurry to try to get rid of the A-10. In 2015 it considered retiring the A-10 fleet as it attempted to squeeze money out of its budget for development of the Lockheed Martin F-35 Lightning II Joint Strike Fighter, another fast mover with what appeared to be limited CAS capability.

The A-10 community, which always considered itself the red-headed stepchild of the Air Force because its mission was not regarded as glamorous by the jet jockeys, protested, and eventually enough money was found to keep the Warthog in the inventory for another decade or so.

After my work on the YA-9 project, I moved on to the F-5E Tiger II, an upgrade of the F-5A and F-5B Freedom Fighter, which Northrop had designed primarily for an air superiority role, but which also had some ground attack capabilities. It was lighter, faster, more maneuverable, and less expensive than the F-4B Phantom, which made it especially attractive to foreign buyers.

At one point in the program, we were having problems with the maneuvering flap on the F-5. (The maneuvering flap allows for tighter turns at high speeds, which is particularly important during air-to-air combat engagements.) If you're flying along straight and level, the flaps stay up. If you start turning and yanking the plane around, the flaps should come down to a pre-set position. The problem was that the system wasn't working that way, and the Northrop bosses were pulling their hair out because we were getting ready to deliver airplanes to the Air Force, and no one could figure out what was wrong.

The Air Force had a lieutenant colonel assigned to the Northrop assembly plant in Palmdale, California. When the airplanes came off the assembly line, he was to sign off on them, and the company would sign off and get a DD-254, which was clearance for the government to pay for the airplane.

Well, this ditzy lieutenant colonel, who was a pilot but not a test pilot, decided one day he was going to take the plane out. I don't know if he thought he could figure out the problem or if he was just screwing around. He certainly was not qualified to test the aircraft, but he took it out, got into trouble, bailed out, and lost the aircraft.

A pilot should always know his limitations and those of the aircraft. This guy obviously did not know either and lost a $2 million aircraft. Obviously, it's better to lose an aircraft than lose a pilot, but this was a case of sheer stupidity.

Not long after that, Hank Choteau came to me and said, "Dave, you had a pretty easy flight today. Can you take this other Tiger out and play around with the maneuvering flap and tell me what's wrong with it?"

I agreed and took off, climbed to 36,000 feet at Mach .9, lit the afterburners, and put the plane straight up vertically. Then, I put it in the maneuvering flap mode. As I started to get a higher angle of attack, the flap started to come down. I got to the top of the climb and ran out of air speed. It was zero air speed.

I said to myself, "Holy shit, Dave! What are you going to do now?"

About that time, the airplane swapped ends and started to yaw. Yaw is bad on airplanes. That's when you start to spin, and spins are bad for the aircraft and are especially bad for the pilot.

I thought, "Damn it! You got into this mess by putting the flaps down. Why not try putting the flaps up and see if you can get out of this spin?"

So, I put the flaps up, and the nose of the airplane tipped over, and I went home.

When I got back to the field, Hank came up to me and said, "Dave, I understand you had a good flight."

I said calmly, "Yeah, it wasn't bad." There was no point in going into a lot of details because Hank had yet to see the flight data, which had been captured back at the base. Hank said the flight engineers were reviewing it to see just what had happened. "When they're done, they want you to go back up and repeat what you just did," he added.

I thought for a minute and then said with a smile, "Hank, why don't you do that? You're more senior than I am."

He just laughed. But we eventually found the problem with the maneuvering flap and it was corrected.

Hank was always kidding me about doing crazy things with the airplanes. He liked to say, "Dave, take this plane up, take a look at it, and let me know what you think." Well, that was an order that meant just about anything. It was up to me to figure out exactly what it was Hank wanted to know about that particular aircraft.

Another time we were having problems igniting the afterburners during high G turns. You want to be able to use the burners during the entire flight, so we had to figure out exactly what was going on.

So, I took the Tiger out and flew about 40 miles northwest of Edwards Air Force Base at 35,000 or 40,000 feet. I rolled it into a tight turn, went firewall with the throttle, lit the burners, and *Poof!* Both engines flamed out.

My first words were, "Oh, hell!" or something to that effect.

I turned back toward the field, called the base, and said, "I just had a dual flameout. I'm going to try to re-start the engines. I'm heading back to the base, but if I don't make it, I'm going to land on a dry lakebed with no power."

Whomever I was talking to turned me over to the air traffic

control tower, and I asked him which was the duty runway at that time. I was coming in from the northwest, and the tower replied that the duty runway was the northern runway.

I responded, "Well, you're going to have to do something. You're going to have to make the duty runway the southern runway."

When he asked why, I told him, "I'm coming in. I have no power. I know I can't make another turn and make it to the north. So, I'm going to land to the south."

"Do you want us to clear the pattern?"

"If you don't, I'll clear it out!"

I kept trying to get the engines started and at 200 feet, I finally got one of the engines lit. But, by then I had the runway in sight, so I could have landed without power even though we were never, ever supposed to make dead-stick landings.

A Northrop flight engineer I told that story to asked me, "Doesn't that danger bother you?"

"Well," I replied, "If it did, I would have to find another job. This is the only job I have right now."

I never told my wife about those sorts of things. When I went home at night and she asked me, "What did you do today, Dave?" I'd say, "I flew airplanes."

One of the things I learned early in my career as a test pilot is that the U. S. government did not particularly care if you lost a $2 million airplane during legitimate testing. My guess is they figured they would rather have the test pilot lose an aircraft as long as the bugs were worked out pre-production so that a normal operations pilot would not encounter a problem and not be able to get out of it. If that happened, he might not just lose the airplane: He might kill himself and hurt people on the ground.

I think the government people believed it was better to *put* the airplane down where there are not a lot of people, get the pilot out, and write off the airplane. My opinion is that's a judgment call the pilot has to make—depending on his ability and limitations. Not all test pilots are created equal. Some are better than others, and some are worse than others. If you get into difficult situations with an airplane you're testing, you must make your own determination about what you're going to do. If you luck out, you live through it, and if you don't luck out, they'll give you a nice funeral.

I think in some respects how far a test pilot pushes the envelope with the airplane is between him and his boss. When you fly, you have a test card that clearly states what the objectives of the test are, and the pilot works to meet those objectives. My boss and I had a really good relationship; he never tried to throttle me and never tried to get me to do more than he thought was necessary.

He would say, "Well, I think Dave will use good judgment and won't unduly sacrifice the airplane or the mission out of stupidity." But, if you never push the boundary a little bit, you'll never know what it's like to fly on all four corners of the flight envelope. (The "four corners" are the aircraft's air speed, weight, altitude, and the gravitational forces, or G-forces, the aircraft can withstand.)

One of my other jobs at Northrop was flight simulator work on what the company called the YF-17 Cobra. It was designed as a smaller, lighter, and less expensive alternative to the Air Force's F-15 Eagle in the service's Lightweight Fighter program.

Hank Choteau took the first flight in the YF-17 on June 9, 1974. According to *Air Force* magazine, when Hank got back on the ground, he was ecstatic about the performance of the plane. "When our designers said that in the YF-17 they were going to give the airplane back to the pilot, they meant it. It's a fighter pilot's fighter," he was quoted as saying.

Despite the high praise, the Air Force decided it did not want the YF-17, settling instead for the General Dynamics F-16 Fighting Falcon. However, the YF-17 became the template for the F/A-18 Hornet, a joint effort by McDonnell Douglas and Northrop to produce an aircraft for the Navy that would complement its F-14 Tomcats. The F-18 was a beefed-up version of the YF-17 with heavier landing gear for duty on aircraft carriers, but it was essentially the same aircraft Northrop had designed.

During my four years at Northrop, my family and I lived in Lancaster, California. Although Lancaster was about 75 miles north of Los Angeles, we were starting to get an overflow of people moving out there who commuted daily to the city. The more Lancaster grew, the more my wife wanted to leave.

Finally, one day she came to me and said, "Dave, I want to get the hell out of California."

I was not particularly fond of California either. I did not want my kids to grow up there; I wanted a more suitable environment for them, something similar to what I had experienced growing up in Michigan.

I also could see potential problems on the horizon for me if I stayed with Northrop. The problems were not with the company or with Hank Choteau, but with who might replace Hank when he retired. During the downturn in the aviation industry a few years earlier, the company laid off one of its test pilots, Dick Thomas. When it resumed hiring, I was hired ahead of Dick. The sense I got was that my F-4 experience was more important to Northrop at that particular time than Dick's F-5 and T-38 experience.

After I had been at Northrop about a year, Dick was re-hired. When that happened, I figured he was going to be Hank Choteau's replacement somewhere down the line. Dick and I never fought, and we never even exchanged harsh words, but he made it clear he resented the fact I had been hired before he had been re-hired, which gave me seniority. Our differences had more to do with him holding a grudge than with me not liking him. He was angry with me even though I had nothing to do with his situation.

I would see Dick in Hank's office all the time, the two of them just shooting the breeze. Finally, I went to Hank and said to him directly, "I understand Dick Thomas is going to be your replacement."

"That's probably true," he replied.

"Well," I said, "I'll start looking for another job."

Hank asked why I would do that, and I told him that Dick and I just did not get along. "If he gets your job, my days with this company will be numbered," I responded.

The uncertainty of what would happen when Hank retired and my desire to get out of California and back to the Midwest convinced me it was time to pull the plug on Northrop, get out of the desert, and head back east.

Kathy's uncle was a Navy lawyer who had a good friend who was a senior executive at the FAA's headquarters in Washington. The uncle put me in touch with the FAA executive who told me they had an opening for a pilot in the agency's Wichita, Kansas, office. He asked if I would be interested in going to Wichita.

Wichita isn't the best place for everybody to live, but it is in the Midwest. I like to say the best way to see Wichita is in your rear-view mirror. Nevertheless, in 1975 I contacted the FAA about the Wichita job and was hired. It was not exactly starting back at the bottom of the FAA barrel, but it was close. Instead of flying high-performance military jets, it was back to flying civilian cats and dogs.

And, despite my misgivings, Dick Thomas never got Hank Choteau's job.

CHAPTER 9
CATS AND DOGS AND
THE FAA, ROUND 2

My move to the FAA's Wichita, Kansas, office in 1975 offered little more than another step up the agency ladder, a bit more money in my paycheck, and a lot of time sitting in an office pushing paper. This stop is not at the forefront of my career memories: I was there, I did that, and I was glad when it was over. Had I stayed in Wichita, I might have spent the rest of my career flying cats and dogs because two of the major cats-and-dogs companies, Beech Aircraft and Cessna, are in Wichita.

Boeing had a factory in Wichita as well, but we had little to do with it because it primarily manufactured parts for large commercial aircraft that it sent to the main assembly plant in Seattle. Learjet also had a facility in Wichita, but my boss, Ron Puckett, did most of the assignments with that company.

Beech and Cessna were both DOA manufacturers, meaning Designated Option Authorization, not Dead-on-Arrival. That meant those two companies had the privilege of signing off on certification of their own aircraft with FAA oversight. Just how much oversight was always something of a moving target, depending on the issue and the manufacturer.

Of course, that DOA acronym was not a favorite among pilots, and anyone who tried to make light of it usually was met with angry stares, as the head of the Central Region for the FAA discovered one night. He had come to Wichita for a meeting of the Society of Experimental Test Pilots (SETP) and apparently got into the booze quite a bit before his speech. "My talk tonight," he began, "is about the DOA program, Dead on Arrival." Given that this was a group of pilots whose jobs were inherently dangerous, his attempt at humor fell flat. Absolutely no one in the audience laughed. There was an uncomfortable silence for a few seconds and then he continued, but I doubt anyone remembered anything he said that night, other than that first ill-advised line.

Since the Wichita job was my second go-around with the FAA, I was a bit higher on the food chain than when I came right out of test pilot school. As a result, I got to see more evidence of just how cozy the relationship was between the big manufacturers and the FAA. It was during those three years that I began to develop my reputation within the agency of being something of a pain in the ass because I started speaking up about what I was seeing and what I thought was not acceptable. (I was not as big a pain in the ass there as I would later become in Atlanta, but I was well on my way.)

The stated *modus operandi* in Wichita was that if a question came up about something to do with certification for a light airplane and there was a dispute involved, the question was pushed up to the Kansas City regional FAA office because it had authority over small airplanes. If there was a question about a big airplane, the FAA office in Seattle handled it.

That was the theory. In reality, it was not like that at all. If there was a dispute over some aspect of certification of a light aircraft, instead of trying to resolve it in either the local Wichita office or the regional Kansas City office, Beech or Cessna or Lear often by-passed those offices and went directly to their Washington, D.C., office. Once the issue got to the Washington office, there was sure to be political pressure brought to bear on the issue, and the outcome almost always favored the manufacturer.

Most of the complaints from the manufacturers did not filter down to my level, although occasionally some did. There was one instance in which my bosses overruled my findings on a project due to what I believe was pressure from the manufacturer or the desire in the local FAA office to avoid the manufacturer pushing the problem to Washington.

This particular project involved an airplane Cessna was trying to get certified for single- pilot rating. I don't recall the exact model of the aircraft, but I was the leader of the FAA team making the evaluation. We flew the airplane a number of hours, and the results were not good. It was obvious to me that it was too much airplane for a single pilot, especially in high-traffic areas.

When the evaluation team sat down and started discussing our findings, the assistant division chief sat in on the meeting. I told him the mistakes made during the flight tests by the single-pilot crew were quite obvious. These mistakes were dumb things that compromised the safety not only of the aircraft in question but also of other aircraft.

"I don't think we ought to consider a single pilot for that airplane," I stated in conclusion.

One of the evaluators on the team was the ex-chief pilot in Washington. He piped up and said, "So-and-so was on that team, and he voted to approve a single pilot, and you with very little experience are saying he doesn't know what he's doing."

"I can't speak for him," I responded. "I can only speak for myself. But that airplane should not be single pilot."

"You're overruled," he barked, cutting off any further discussion.

"Fine," I replied, even though I was convinced it was too much airplane for a single pilot to handle. The evaluators did not even consider what I had found in my evaluation. And, for him to say I had very little experience was an insult and an indication to me that he just wanted to be finished with the project and was not especially concerned about the safety of the aircraft or its pilot.

Occasionally, I was given projects that made little sense as to why I was chosen as the test pilot. One in particular involved an old Sikorsky H-34 helicopter that was being converted from military to civilian use. In order to certify that it met civilian requirements, it had to go through a production flight test.

I was in Kansas City on another project when my boss called and said he wanted me to fly the test on this old helicopter.

"I can't do that," I answered. "I'm not helicopter qualified."

"It'll be OK," he responded. "This pilot has more time in helicopters than you have in airplanes. In fact, he was a pilot at Wounded Knee and had two helicopters shot out from under him. He'll take care of you."

The "Wounded Knee" to which my boss referred was the South Dakota town occupied for 71 days in 1973 by the American Indian Movement and their supporters in an effort to focus international attention on long-standing Native American grievances. It also was the scene of a controversial 1890 massacre of 150 Native Americans by U.S. soldiers.

Although there is no historical record of any helicopters being shot down during the 1973 siege, helicopters and other military assets *were* used by federal marshals and the FBI in an effort to dislodge the protestors. I am not sure whether my boss added that comment to make me believe the pilot was better than he actually was or whether the pilot had told him that in an effort to bolster his own credentials.

After I got to the airfield and found the pilot, he took me out to the helicopter, and we began walking around it, completing our pre-flight inspection. As we were doing that, I got the strange feeling he thought I actually knew something about flying helicopters.

As I have said before, I never really had any desire to learn how to fly helicopters. I had flown *in* a helicopter once in test pilot school as a non-flying crewmember. It only further convinced me that it would be far easier to take a helicopter pilot and teach him how to fly a fixed-wing aircraft than to take a fixed-wing pilot and make him a helicopter pilot. (My son, David, made that transition from rotary-wing to fixed-wing while he was in the Navy years later.)

When you're flying a fixed-wing aircraft, after a while what you do is instinctive. In a helicopter, nothing you do is instinctive. You do some of the dumbest things for the dumbest reasons, but, in fact, it's the right thing to do. It's not something you learn easily.

After we finished the walk around the helicopter and prepared to board, the pilot said, "Go ahead and jump in the right seat." (In a helicopter, that is the pilot's seat, which is the reverse in a fixed-wing aircraft.)

I responded, "No, I'll take the left seat."

"No," he insisted, "you can have the right seat." He sensed my reluctance and finally asked me, "You are qualified in this aircraft, aren't you?"

"No."

"You are qualified in helicopters, aren't you?"

"No."

"How much helicopter time do you have?"

"When we get airborne it will be my first."

He was quiet for a minute, looked at me with a wry smile and responded, "You're right. I better take the right seat."

I wasn't qualified to fly the helicopter, but my boss assumed I knew enough about flying that I could follow a checklist as well as the pilot. So, the pilot did most of the flying while I handled the checklist for certification — although I flew a bit and got fairly good at hovering and all those other strange things you can do with a helicopter.

The bottom line is you don't have to be a rocket scientist to conduct a production flight test if you know anything about aviation. I observed what the pilot was doing to see if he was doing it correctly, even if I could not do it myself.

Another of the more interesting projects in which I was involved was with a twin-engine jet owned by a Swiss doctor who wanted the plane to get approval for a Category II (CAT II) landing. That is an instrument approach, usually in low-visibility conditions, in which the decision to land or abort is made between 100 and 200 feet above the runway. In CAT I the decision height is no lower than 200 feet while CAT III has three categories, ranging from the decision height lower than 100 feet to automatic landing and rollout.

Getting a CAT II approval is a big deal because you must perform a lot of flight tests at different airports that are qualified for CAT II. Not only are there certain standards for the aircraft, but the pilots must also meet specific standards to fly that aircraft. And, once a CAT II rating is acquired, it is not grandfathered in for life; it has to be maintained through regular testing.

The airplane this doctor wanted certified for CAT II was something like a Gulfstream, although it was not a Gulfstream. The plane had a professional crew that regularly flew it. My responsibility was to test the plane and crew to make sure they met the CAT II landing requirements. (The project was actually being performed under the direction of King Radio, which did a considerable amount of avionics work in that area, and one of its engineers usually rode along on the test flights.)

The Swiss pilots for this particular aircraft were having some problems with their landings on the 5,000-foot runway, so I took the plane up and executed a modified carrier landing, touching down near the beginning of the runway and turning off at about 2,500 feet.

When I did that, the chief engineer for King Radio, who was sitting in the first passenger seat behind the cockpit, broke out laughing. I asked him what he found so amusing.

"Do you know why those pilots have problems landing on this runway?"

"No," I replied, "I don't understand it."

"They don't land until they are halfway down the runway. You turned off the runway where they usually land."

The Swiss pilots apparently were used to coming in high and landing well down the runway, unlike me, who was used to landing early because of my days of carrier landings. Carrier pilots never land late.

"Well," I chuckled, "the runway behind you never does you any good!"

In aviation, the three most useless things are the runway behind you, the altitude above you, and the gasoline you left at home.

As I recall, that aircraft never was certified for CAT II landings.

Wichita was where I first began pushing back against the way the FAA conducted some of its business. I think pushing back was a learned trait. Somewhere along the line, everybody has to make a judgment call as to what they believe is right and what they believe is wrong, and what they are willing to stand up for and what they're not willing to stand up for.

I think that trait was born in me. Or, maybe that's my father's influence again. The key, though, is doing the right thing when nobody is watching. You can go on the six o'clock news and say all you want about how you did things, but it's what you do when the cameras aren't rolling that counts.

I left Wichita in 1978 and went to the FAA regional office in Kansas City for a promotion from GS-13 to GS-14. The Kansas City job was an office job, but it was a promotion and meant more money. I knew I wasn't going anywhere if I stayed in Wichita. As a staff member in Kansas City, I pushed a lot of paper. My primary job was to review findings from different offices in our region. Occasionally, I went out into the field and made some findings myself, and most of that involved small airplanes. I still got to fly, but at that time the paychecks were more important than flying. The work was actually rather boring.

When the opportunity came in 1982 to move to the FAA's Atlanta office, I knew there would be more opportunities there than I had in Kansas City. The only issue standing in the way was that my older son, David, was about to start his senior year in high school, and my wife was dead set against making him move and leaving his friends behind. He was only nine when we moved to Kansas, so his formative teen years had been spent in the Midwest.

I remember one morning not long after we had moved to Wichita from the California desert that David woke up, walked into the kitchen, and said, "It's not so hot here but there's something wrong with the heat."

What was wrong with the heat in Wichita was the humidity. There wasn't any of it in the high, dry desert country of Lancaster, California. David knew the air was different; he just did not know what to call it. If David thought it was humid in Wichita, he did not know what was in store for him when he got to Atlanta.

My wife suggested David could stay in Kansas City with some friends who had a son the same age and finish out his senior year before going to college. I did not like that idea and decided not to allow her to make it an issue.

"No, he's going with us," I said with finality. "Case closed."

The wandering Wests were off again, this time headed east for my job in the FAA's Atlanta office. My seniority in the agency had increased to the point where I became even more outspoken about things that I thought were not being done in the best interests of the flying public. One of my first projects in Atlanta convinced me to push back against the entire hierarchy of the FAA because of what I believed was a poorly designed system that — if installed in commercial passenger aircraft — could have cost hundreds of lives.

It was a decision on my part that earned me the enmity of a number of fellow employees and the reputation of being a big-mouth troublemaker in an agency that prized a "go along to get along" attitude among its employees. Doing what I believed was the right thing at the right time was the only way to do it, even if it meant the possibility of losing my job.

CHAPTER 10
ATLANTA, THE TCAS FIASCO,
AND THE FAA, ROUND THREE[1]

On October 29, 2018, Lion Air flight 610 took off from Doekar-no-Hatta International Airport in Jakarta, Indonesia, with 189 passengers and crew. The plane was only 13 minutes into its scheduled 75-minute flight to Pangkal Pinang, Indonesia, when it crashed into the Java Sea, killing all on board.

Less than five months later, Ethiopian Airlines flight 302 departed Bole International Airport in Addis Ababa, Ethiopia, with 157 passengers and crew. Just six minutes into its scheduled two-hour flight to Nairobi, Kenya, the plane nose-dived into the ground. There were no survivors.

After nearly five decades in aviation, anything related to flying or aircraft is of great interest to me. I read about the Lion Air crash with some curiosity because the aircraft was a relatively new Boeing 737 MAX. The Ethiopian Airlines crash further piqued

---

1. Some of the information in this chapter became public only after the death of the author. It was obtained from court documents, newspaper articles, and other open-source materials.

my interest because it involved another 737 MAX. You don't have to be a rocket scientist—or even a test pilot—to figure out that when two aircraft of the same model and the same recent vintage go down so close together that there just may be something wrong with that model.

Over the next few months, as I read numerous news articles about the causes of these crashes, it became obvious to me the responsibility for them lay at the feet of both Boeing and the FAA. Published reports clearly showed a rush by Boeing to produce the 737 MAX—because of competitive pressures—in addition to an egregious lack of safety oversight by the FAA. The result was that those planes were allowed to fly before they had been fully tested and certified as airworthy, and 346 people paid with their lives because of the lack of adequate oversight.

The 737 MAX crashes and subsequent grounding of that model aircraft worldwide brought back a wealth of bad memories of my time with the FAA's Atlanta office and a remarkably similar situation which involved me as the chief test pilot. That specific project nearly cost me my job and did nothing to endear me to the FAA hierarchy because of my insistence that safety—not dollars or deadlines—was our top priority for the flying public.

In the spring of 1982, the position as chief of flight test came open in both the Wichita and Atlanta offices. I bid on both jobs and thankfully got the Atlanta job. Had I gotten the Wichita job, I do not know how long I would have lasted there because it would have been more aircraft cats and dogs for the foreseeable future, if not for the rest of my career.

In addition to being a promotion from the GS-14 position to GS-15, the Atlanta job was considered a plum posting within the FAA because it offered many more opportunities and much more responsibility. At that time, the Atlanta office had certification responsibility not only for aircraft in the southeastern United States, but also for the Caribbean and all of South America, which included an Embraer factory in Brazil that operated under our guidance. Over the next few years, I made a number of trips to Brazil; that never would have happened had I gotten the job in Wichita.

I moved the family to the Brookfield West golf community in Roswell, Georgia, north of Atlanta, in August 1982. Our house was about 35 miles one way and at least an hour or more commute in miserable traffic to the FAA office in College Park near the Atlanta airport. We chose to live so far from my office because my wife had done some research and had determined Roswell High School was one of the top schools in the state. Since I insisted our older son, David, could not stay in Kansas City to complete his senior year of high school, my wife insisted we get him into a good school in Georgia. David graduated from Roswell High School in 1983 and went on to the U.S. Naval Academy, following in his old man's footsteps.

I've always been practical, whether it was in my personal life or in the aviation industry. Getting right to the point of whatever needed to be done and getting it done in the fastest, most economical, and safest way possible was how I liked to do things. But, it had to be done correctly or not at all, as far as I was concerned. I brought that sort of attitude to the FAA's Atlanta office, where we handled a great deal of aviation business. For some people who worked there, my method seemed to be a novel way of doing things, and I was not the most popular guy in the office at times, especially with management.

The southeastern region of the FAA includes Alabama, Florida, Georgia, Mississippi, North Carolina, and South Carolina. Piper Aircraft was in Vero Beach, Florida. Gulfstream Aerospace Corporation was in Savannah, Georgia. Alabama and Orlando,

Florida, had major repair facilities with which we worked quite a bit. TIMCO Aviation Services in Greensboro, North Carolina, did a number of modifications on large airplanes, and I spent a great deal of time there. In the Caribbean, we did a lot of cats and dogs, checking on various modifiers, such as hush kits to muffle engines on a variety of aircraft.

The Atlanta FAA office probably had about 50 people, including clerical staff, working there in the early 1980s. We had subject matter experts in systems, propulsion, airframes, manufacturing inspection, and flight test, with a substantial number of projects going on at any one time.

Probably the most significant and most frustrating project on which I worked while in Atlanta was TCAS (pronounced TEE-kass), which stands for Traffic Collision Avoidance System. TCAS had been in development for nearly 30 years by the time I got to Atlanta, but the fine-tuning of computer algorithms and deconfliction of radio and transponder signals that would make the system work properly was proving to be a headache for the designers. That was especially true for aircraft trying to use the initial versions of the system in high-density airspace such as those around major airports in Atlanta, New York, and Los Angeles.

The impetus for development of a collision avoidance system was a 1956 mid-air crash involving a United Airlines DC-7 and a Trans World Airlines L-1049 Super Constellation over the Grand Canyon. All 128 passengers and crew on the two planes were killed. It was the first time more than 100 lives had been lost in a commercial airline crash.

The intent of the TCAS program was to equip aircraft with a system that would enable pilots to detect nearby aircraft independent of the ground-based air traffic control system and to make the necessary corrections to avoid a collision. There was even more of a push to get that system up and running after two more major mid-air collisions over the next 20 years. In 1960, a United Airlines DC-8 and a TWA L-1049 collided over New York City, killing 128 passengers and crew and six people on the ground. In 1978, a Pacific Southwest Airways 727 collided with a Cessna 172 over San Diego, California, killing 137 passengers and crew on the two aircraft and seven people on the ground. It was the worst aviation disaster in history at that time.

As commercial air travel increased in popularity and became more economically viable for large numbers of people, Congress put increasing pressure on the FAA to come up with a solution that would greatly reduce, if not eliminate, these types of aviation catastrophes. TCAS was supposed to be a system that made flying safe for everyone, including nitwits.

The TCAS development in the late 1970s and early 1980s took two concurrent tracks. TCAS I was designed for turbine-powered passenger aircraft with more than 10 seats and fewer than 31 seats. It was considered a cheaper alternative to TCAS II and provided the pilot only with what are known as TAs, or traffic alerts, which meant the system told the pilot only that there were other aircraft nearby; it provided no information about the altitude or speed or direction of those other aircraft. It was the pilot's responsibility to visually acquire and maneuver away from the potential threat or threats.

TCAS II was designed for commercial aircraft with more than 30 seats and a maximum takeoff weight of more than 33,000 pounds. It eventually provided the pilot with both TAs and RAs, or resolution alerts, the latter telling the pilot how to maneuver to avoid a collision. The problem with the early TCAS II system on which I worked was that the RAs did not do what was originally intended: TCAS II simply told the pilot to ascend or descend; it could not tell the pilot to turn left or right, which created a whole host of problems that, much to the chagrin of my bosses, I kept pointing out to them as being a dangerous shortcoming in the system.

Atlanta had responsibility for the TCAS II testing because the first systems were put on two Piedmont Airlines Boeing 727s based in Charlotte, North Carolina. Although the system was supposed to be designed primarily for Instrument Flight Rules (IFR) flying, we never tested the early models under those conditions. That failure should have clued someone in right then that there was a major problem.

On all those early test flights with TCAS II equipment in which I was the primary test pilot, I never saw any of the data generated by the computers as to whether the system worked as it was supposed to. Flight engineers monitored the computers out of sight of the flight crew, so it was impossible for me to tell whether it was working properly. And, again, we only did the testing under visual flight rules (VFR), never under IFR.

My big concern with the system was the fact that early versions of TCAS II did not actually tell the pilot what to do, only that he should climb or descend. But, once it told the pilot to climb or descend, it never changed its mind until at the very end it would say, basically, "I give up. You're on your own." If you're flying IFR and the computer tells you that, how are you supposed to know what the hell to do?

I went to my bosses and said, "That's unacceptable."

"Why?" they asked.

I explained as clearly and patiently as I was able that it's quite common for airplanes coming into Atlanta from the west to descend before they make the turn to go around and land to the west. Everyone knows that air traffic control is going to level that incoming flight at several thousand feet below the outgoing flight. The outgoing flight then passes over the incoming flight.

TCAS II did not know that, though, and thought the two planes were going to crash, so it would tell the outbound flight, "Descend! Descend!" But, if that airplane descended, it was going to collide with the incoming flight. That same traffic pattern is typical at most major airports. TCAS II as it was originally designed could literally drive a plane into the ground. It could tell a pilot to keep descending until there was no more descent left, and the plane *was* planted in the runway or a nearby field.

One of the other questions I asked was, "Why can't TCAS II change its mind when it sees something else happening that it wasn't aware of before?" Once it issued an RA, it could not change, even if new information came in.

My bosses just said, "Oh, that's a software problem. We'd have to change the software."

I said, "Well, change the software."

They did not want to do that, though, because doing so would have taken more time and cost much more money, and money was a key consideration whenever significant changes were made. They wanted to get the system certified. They wanted to go before Congress and say, "Here we are…we got your system for you." But they did not want to spend the money or the time to do it correctly.

The Congressional mandate to approve TCAS II was a known fact. Every time I brought up problems with the system I was told, "We don't have time to make changes."

When I asked why not, I was told, "Because Congress has mandated it be done by such-and-such a date, and we're going to get it done."

If there was a specific date by which the system had to be certified, I was never given that information. We had numerous meetings about implementation of the TCAS II system. Sometimes it would be a teleconference, sometimes the Washington people would come to Atlanta, and sometimes we would go to Washington. We always had our bags packed to head to Washington with very little notice.

In almost every meeting the Washington people let us know they wanted to certify the system no matter its shortcomings. Many of the meetings we had over the certification of TCAS II were heated, with the bosses pushing for its approval and me pushing back because I did not believe the software could do what it was supposed to do. The only ally I had in the fight was an engineer in the Seattle office who for her sake I will refer to only as "Sarah."

Sarah and I told the Washington people they didn't know what they were doing. We told them they were going to get people killed unless they worked out the bugs in the system before they certified it for use on commercial airliners, and we argued until we were blue in the face.

The No. 2 man in aircraft certification for the FAA at that time was Tom McSweeney, an aeronautical engineer who was at Northrop about the same time as I and had left Northrop to work for the FAA in the Los Angeles office. McSweeney was absolutely adamant we were going to approve that system, despite its flaws.

After one particularly contentious meeting via teleconference, I turned to McSweeney and said, "Tom, do you have a problem looking at yourself in the mirror in the morning when you shave?"

"No," he replied. "Why should I?"

"I don't understand how you can look at yourself in the mirror because you're coming up with some of the most irrational arguments I've ever heard in my life. What you're trying to do will get people killed!"

He just mumbled a little, but my boss, John James, an old World War II bomber pilot, put his head in his hands and went, "Oh, no." I thought for a minute John was going to die of embarrassment. I've never been one to hold back when I have an opinion on something, and I certainly was not about to hold back in this case because the flying public's safety was at stake.

Management knew they could not intimidate me into backing down because I already had a reputation of not going along to get along just because the bosses wanted me to. Since they could not get to me, I got the sense that they believed they could intimidate Sarah just because she was a woman.

Another of the FAA certification officials in the Washington office who was pushing to get TCAS II certified was Tony Broderick. Tony was not a pilot. Prior to joining the FAA, he had worked on ozone reduction for the Department of Transportation. One day in one of our teleconferences, he made some sort of smart-ass remark to Sarah, and she bristled.

"Tony," she said sharply, "are you trying to intimidate me?"

Broderick quickly denied it, but just about everybody on that teleconference knew that was exactly what he was trying to do. He was trying to intimidate her into coming over to his side so they would then have only me to deal with.

I was not about to back down because I was approaching the TCAS II certification from a technical standpoint; the FAA bureaucracy was approaching it from political and financial standpoints. The agency has the responsibility to oversee the manufacturers, but often it does not take that responsibility. The FAA's answer was always, "The manufacturer knows what he's doing."

Well, I've often thought, if that's the case, why do we have the FAA?

I revisited that question about the FAA's purpose numerous

times years later as I read about the investigations into the crashes of the two 737 MAX aircraft. The situation I encountered with TCAS II seemed to have gotten worse over the years.

The preliminary investigative findings into the 737 MAX crashes by the House Committee on Transportation and Infrastructure issued in March 2020 only confirmed my suspicions about how bad things had gotten in the FAA and how it dealt with one of the major aircraft manufacturers.

According to the House report, the simple answer for the 737 MAX crashes was a new system known as MCAS (pronounced EM-kass), which stands for Maneuvering Characteristics Augmentation System. MCAS was designed to provide longitudinal stability for the 737 MAX. Longitudinal stability, sometimes referred to as pitch stability, is the aircraft's ability to avoid diving or climbing as a result of the nose moving up and down during flight. If an aircraft is longitudinally unstable, the up or down motion could send it into a steep dive or a climb, resulting in a stall.

Designers found that due to the larger engines on the 737 MAX, an abnormal nose-up pitch developed on takeoff at a high angle of attack (AOA). MCAS was installed to gradually push the jet's nose down to prevent stalls when the flaps are up and there is a high angle of attack.

If there was a disagreement between the two AOA sensors on either side of the aircraft's nose, a warning light was supposed to come on in the cockpit. It apparently did not in either of the 737 MAX crashes and the MCAS, sensing the AOA was too steep and the two planes were about to stall, took over and started pushing the noses down. The pilots apparently were unable to disengage the system. Essentially, the computer took over and drove those planes down, one into the ground, the other into the sea.

In its preliminary report, the House committee found six major areas of concern with Boeing, the FAA, and MCAS. Quoting the report here, investigators found:

- "The FAA failed in its oversight responsibilities to ensure the safety of the traveling public."
- "Costs, schedule, and production pressures at Boeing undermined safety of the 737 MAX."
- "Boeing failed to appropriately classify MCAS as a safety-critical system, concealed critical information about MCAS from pilots, and sought to diminish focus on MCAS as a 'new system' in order to avoid greater FAA scrutiny and increased pilot training requirements."
- "Boeing intentionally concealed information from the FAA, its customers, and pilots about inoperable AOA Disagree alerts installed on most of the 737 MAX fleet, despite their functioning being 'mandatory' on all 737 MAX aircraft, and the FAA has failed to hold Boeing accountable for these actions."
- "Boeing's economic incentives affected the company's transparency with the FAA, customers, and 737 MAX pilots regarding pilot training requirements."
- "Both Boeing and the FAA gambled with the public's safety in the aftermath of the Lion Air crash, resulting in the death of 157 more individuals on Ethiopian Airlines Flight 302 less than five months later."

In January 2021, Boeing agreed to pay a $2.5 billion settlement to avoid criminal prosecution after the U.S. Justice Department charged the aircraft maker with fraud regarding the bungled certification of the 737 MAX.

Included in that settlement was a fine of $243.6 million plus

$1.77 billion that went to the airlines that had ordered the planes but were left holding the bag when the 737 MAX was grounded for more than two years.

In addition, the families of the victims of the two crashes shared a settlement of $500 million, or about $1.45 million per family, as long as they submitted their claims properly and in a timely fashion. That did not preclude civil actions against Boeing.

As might be expected, there was a shakeup in the Boeing hierarchy as a result of the crashes and subsequent investigations. Some people were fired, and some took early retirement. Nevertheless, nobody went to jail for any of the omissions the company perpetrated in developing the plane. Nor was anyone punished for the deaths or the efforts to conceal any of the evidence that this plane was not ready to fly when it was certified.

The only person charged with a crime was Mark Forkner, the chief technical pilot for the 737 MAX during its development at Boeing. A federal grand jury in Texas indicted Forkner on six counts of fraud. Forkner left Boeing in 2018 and went to work for Dallas-based Southwest Airlines, which was the biggest customer of the 737 MAX. The company already had 34 of the planes in its fleet and hundreds more were on order.

Forkner went on trial in Fort Worth, Texas, in March 2022. After four days of technical testimony, a jury took less than two hours to acquit him on all charges. He had come across as the villain in the story of the 737 MAX because of a series of profane text messages he had exchanged with a co-worker. In one of those, he wrote, "This airplane is designed by clowns, who are in turn supervised by monkeys."

In another message, he wrote that a Boeing presentation to FAA officials on the 737 MAX was so complicated that "it was like dogs watching TV."

The Congressional hearings and the trial revealed a great deal about the inner workings of Boeing and pressures brought to bear on its employees to get the 737 MAX certified because of what was at stake financially.

The hearings also revealed that the crux of the problem was in the MCAS software, something for which the engineers had been responsible. The engineers had changed the software to allow MCAS to activate at low speeds as well as high speeds. So, a single faulty angle of attack sensor could activate the system at lower speeds, something for which the pilots in the two crashes apparently had not been told about or trained to recognize.

The more I read about MCAS and how the FAA abrogated its responsibility to Boeing for certifying the 737 MAX and how Boeing failed in its responsibility to the flying public, the more it sounded like what we had gone through with TCAS II. The only difference was that with TCAS II we were successful in getting the software changed so that it provided the types of RAs that I thought made it a much safer and much more dependable system. Somebody must have gotten to the bureaucrats at some point during the TCAS II development and said to them, "These people have concerns, and they seem to be legitimate concerns. Why aren't you listening?"

TCAS II was finally certified in April 1986. To mark the occasion, the FAA flew everybody who had worked on the project to Washington to celebrate, with two exceptions: Sarah and I were left off the guest list. I am sure it had to do with us not being what the FAA bureaucracy considered team players. At the end of that particular project there were not too many people in FAA certification who did not know who Dave West was.

About a week after the certification celebration, a package arrived in the mail for me. When I opened it, inside was a plaque thanking me for my outstanding performance on TCAS II. I was so disgusted with the whole thing that I threw that plaque away so fast it would have made your head spin. There was no way I was going to put that thing on my wall.

The TCAS II and the MCAS projects raise some significant issues about the FAA certification process. Just who *is* running the show? In my opinion, who runs the show depends on who you are. If you're Boeing or Cessna or Beech, you get pretty much what you want. If you're some little guy out there with a little airplane that you want modified, they hammer you to death with oversight, evaluations, and more paperwork.

I had some real heartburn with the FAA and the way it did business. Of course, I think I gave a lot of people in the FAA heartburn because of my insistence on following the rules and doing things the right way.

A case in point was the issue of training for FAA test pilots. At that time there was an FAA directive that clearly stated that every test pilot must have recurrent training in some aircraft at least once a year. It did not recommend it; the directive said it *will* happen. However, we were not getting that training in the Atlanta ACO (Aircraft Certification Office). At that time, all aircraft certification officers were attached to regional offices in Kansas City, Boston, Dallas, or Los Angeles, so as a certification officer in Atlanta, I was under the Kansas City regional office.

The director of the Kansas City office came to Atlanta one day and I decided to confront him about the training issue. I picked up a copy of the directive and said, "Director, may I have a word with you?"

"Sure," he said.

"We're having a problem."

"What's the problem, Dave?"

"We have an order that says certification test pilots will get annual recurrent training in an aircraft."

"What order?"

I showed him the directive and he read it. Then, he looked at me and said, "Oh, that's not required."

"It's not?"

"No, it's not required. That's just nice if we're able to do it."

"Where do you see that? It says you *will* have the training."

"It's just a suggestion."

"Do you mean to tell me that this order is just a suggestion, that I can follow it if I want to but don't really have to?"

"You have to follow orders," he responded.

"Well, why don't we have to follow this one?"

"You're just confused."

And I shot back, "I think *you're* confused." He looked at me with consternation, but I was not to be ignored.

My persistence paid off, and all of the pilots in the Atlanta ACO received their required recurrent training in that particular year. Recurrent training was—and still is—essential for flight safety. Just ask any pilot what he or she thinks about the issue of recurrent training.

I could get away with things like that because I never really worried about my job. As a test pilot, I knew I could get another test pilot job in the private sector the next day if I were fired. I am not sure how close I ever came to being fired, but I sure pressed hard against my boss at times. A good number of those times were with a subsequent boss in the Atlanta office, a man I will refer to only as "John." He was a complete idiot and proved it almost daily.

On one occasion, we had a meeting in Atlanta with branch managers about evaluating employee performance. There was a sense that bosses were grading their personnel too high and that the grades needed to be lowered. John asked every branch manager to stand up and justify the scores they gave their people. One manager got up and said he had a secretary whose elevator did not go all the way to the top floor but that he had given her an "Outstanding."

When the branch manager had finished, John turned to me and said, "Well, Dave, what do you think?"

"Well," I said, "if she worked for me, she would have gotten an 'Unsatisfactory.'"

He shook his head and looked at me as if I was a fool. "You don't understand the promotion system in the FAA," he commented.

"Probably not," I said. "Why don't you explain it to me?"

He got this look on his face as if he thought it beneath him to have to explain this to me, let out a sigh of exasperation, and then responded, "Sometimes, when you have an employee who is doing less than average work you give them an 'Outstanding' to raise their capabilities."

I looked at him with some amusement and asked, "Well then, John, why didn't you give me an 'Outstanding'?"

"Why? What do you mean?"

"Because you're always telling me I don't know what the hell I'm doing." That got a laugh from everyone in the room but John.

Whatever that system was that John was talking about was not a ratings system.

John and I also disagreed about how best to utilize test pilots assigned to the Atlanta office. He did not want his test pilots out flying; instead, he wanted them in the office pushing paper. I did what I could to get out of the office whenever possible, though. In one year, I had more than 175,000 miles flying to assignments on Delta Airlines alone. That does not include flights I took on other major airlines. I also occasionally rented airplanes and flew to an assignment if it was close to home.

One of my biggest disagreements with John was over an issue related to Gulfstream Aerospace, a manufacturer of small business jets based in Savannah, Georgia. The company is just a few hours down the road from Atlanta, and one of our better pilots, Bob Sample, worked closely with the company on several projects.

Sample was a fellow naval aviator and had been commanding officer of the U.S. Navy Test Pilot School at Patuxent River, Maryland, before joining the FAA. His work with Gulfstream was so outstanding the company commended him in letters to our office, citing his professionalism and skills as a pilot. At one point, though, an issue arose over which Bob and Gulfstream were at odds. Bob would not back down from whatever change he requested, so Gulfstream officials wrote a letter to John complaining about how unprofessional and uncooperative Bob was being.

John called me into his office after he got the letter from Gulfstream and asked me what I thought about it. I pulled out the letter we had received from Gulfstream just a few months earlier praising Bob's performance, showed it to John, and asked, "Which one of these is true?"

John indicated the latest one, not the earlier letters, must be true, which made absolutely no sense to me.

"I don't think so," I countered.

John said he wanted me to write a letter of reprimand for Bob's official FAA file. I refused. I told John there was no way I was going to write that letter of reprimand because I knew that if Bob thought what he was doing was the right thing, I was going to back him. Bob and I clearly shared the same philosophy about doing things the right way.

The letter of reprimand never was written, but John took Bob

Sample off any future work with Gulfstream. John also never talked to me about it again because he did not have the guts to take me on. Not long after that episode, I was in my office when I overheard a conversation coming from the next one, John's office. One voice was that of the Central Division chief who yelled, "Well, goddammit! Who does Dave West think he is that he thinks he's going to tell us how to run the aviation industry! Maybe we ought to get rid of him."

The division chief was not big enough to take me on either, and while I thought for a moment or two I might lose my job, I would have gone that route if it meant standing up for one of my pilots for doing what he thought was the right thing to do.

John left the Atlanta office not long after that and went to Kansas City before returning to Georgia and going to work for — not surprisingly — Gulfstream. It was not unusual for FAA executives to go to work for major aircraft manufacturers, usually for much more money; after all, many of those FAA officials had worked closely with those companies for years, and hiring someone from inside the FAA was a way for companies to exercise the kind of influence with the agency that was invaluable.

Some time after John left, I was heading out to lunch, and just as I got to the front door of the building, in walked John. He had this big, shit-eating grin on his face and held out his hand to me. I refused to shake his hand. Instead, I just looked at him and exclaimed, "I'll be damned! Look what comes into view when you don't have a shotgun!"

By 1985 it was clear the job with FAA had taken precedence over my family life. One day my wife, Kathy, came to me and said she wanted a divorce. I did not argue and gave her what she wanted. Before we separated, though, she wanted an answer to a question that had been bothering her for years. I had attended a number of schools years earlier in which I picked up a great deal of classified information, much of it considered "Top Secret," which I never discussed with her.

"Dave," she said one day, "I know you're a spy. I just don't know who for."

"You'll never know," I replied with a smug smile. And...she still does not know.

I now consider my divorce and World War II in the same category: ancient history with little further to be discussed.

A year after the divorce, I met Becky. She was living in Fayetteville, Georgia, with her two children following her divorce in Maryland in 1985. She was certified to teach in Virginia and West Virginia, but Georgia would not recognize those certificates, so she went back to college part-time to get a Georgia teaching certificate. She drove to classes at West Georgia College (now the University of West Georgia) in Carrollton, Georgia, with a woman I had dated several times following my divorce. The lady and I never really hit it off, but she thought Becky and I might be a better match, so she gave me Becky's telephone number.

I held on to the number for three weeks before calling Becky and asking her out to dinner with Bob Sample and his wife, Joan. I told Becky to bring a swimsuit along because the Samples had a pool, and we planned to go for a swim after dinner.

While Becky was standing on the edge of the pool talking to Bob, I sneaked up behind her and pushed her in. It may seem a bit juvenile to have done it, especially on a first date, but I later told her that it had been a test: If she came out of the water laughing, I knew she was a good egg and somebody I would see again. If she came up pissed off, I would never call her again. She came up laughing, so she passed the test. A year later, in 1987, we were married.

After the wedding, it was back to work on numerous assign-ments over the next few years that kept me on the road and in the air for weeks at a time. In addition to stateside jobs, there were trips to Brazil to work on Embraer projects, a rather strange trip to Switzerland to fly an airplane that hardly flew, and some par-ticularly memorable trips to China, one of which involved a plane that used bicycle chains to raise and lower the flaps. And it was on one of those China trips that my former wife's belief that I was a spy nearly came true.

CHAPTER 11
SPIES, LIES, AND BICYCLE CHAINS

The China Expeditions and Other FAA Follies

The life of a civilian test pilot, especially one who works for the FAA, is not always as glamorous or as exciting as it might seem. For every high-speed, low-drag jet I got to fly, there were any number of flights to check out a minor modification on an airplane to make sure it met safety standards. During my nearly 15 years with the FAA office in Atlanta, I did test flights on everything from jumbo passenger jets to hot air balloons.

That hot air balloon test was in South Carolina in the early 1990s. There is not really much testing you can do on a hot air balloon: It goes up and it comes down. The instrumentation is rudimentary because a balloon is not a sophisticated piece of equipment, and all you're really looking for are safety issues. But, when I got to the test site in South Carolina and saw the balloon, I did not know what to think: It was shaped like a giant condom. (This was at a time when the AIDS epidemic was still a world-wide problem, and the balloon was designed to promote a particular condom manufacturer.)

I no longer recall exactly what I tested on that balloon because I could not get that image of a giant condom floating through the air out of my mind. It still creeps back in occasionally, often at the most inopportune times. When I returned to Atlanta and told my co-workers about it, they dubbed me the FAA's "Chief Condom-naut" and created a special nameplate for my desk.

Another less-than-glamorous test was in Switzerland. I was sent there to check on a twin-engine aircraft the manufacturer wanted to certify for use in the United States. The company's chief pilot was Vietnamese and did not say much about the plane prior to the test. One day I took the plane out while he sat in the co-pilot's seat. As we were flying along, I commented to him, "This airplane has no lateral directional stability." (That meant the pilot had to continually fight the aircraft to keep it from rolling.)

"I know," he replied.

We flew a little bit more and I said, "This airplane has no longitudinal stability." (That meant the pilot had to fight the controls to keep flying level.)

"I know that, too," he responded.

At that point, I thought to myself, "Why in the hell am I here?"

It was a horrible airplane, and I refused to certify it. Eventually, it was certified to fly in the U.S., but that certification did not come from Dave West. If I did not believe an aircraft was airworthy, I refused to certify it, no matter how much pressure my bosses put on me to do so.

The FAA is part of the federal government bureaucracy and the best way to get along in a bureaucracy is to keep your mouth shut and do what your bosses tell you to do. Don't make waves. Don't make anybody angry. Go along to get along. That attitude was not part of my makeup, though. If something was not done the right way for the right reasons at the right time, I was likely to raise some hell about it, just as I did during the development of TCAS II and with John and the Gulfstream issue. It did not even have to be something major.

A case in point was a trip Bob Sample and I took to Wichita for a project in the 1980s, the details of which are buried somewhere deep in my logbook. What was significant about that trip, and the reason it sticks in my memory, occurred as we were preparing to leave the Kansas City airport for the short flight to Wichita on a commuter airplane.

This particular plane belonged to a small Wichita-based aviation company and had no overhead luggage compartments, so we had to hold our bags on our laps. Bob and I were in the two front seats, and a few minutes before we were scheduled to leave, a young woman came up the ladder. "Well, there's the flight attendant," I said to Bob. "We'll be under way soon now."

She was not the flight attendant, though; she jumped into the co-pilot's seat. As soon as she was seated, the pilot cranked up the engines, and we began taxiing to the runway. Then he revved up the engines and started down the runway.

The plane went only about 100 feet when the pilot throttled the engines back to idle, picked up the intercom, and announced, "Ladies and gentlemen, I'm sorry, but I attempted to taxi with the parking brake on."

I just shook my head. This pilot did not have his act together, and his co-pilot was no help. While we waited for the brakes to cool off so we could make another attempt to get off the ground, I wondered to myself, "How long has this guy been flying?"

Bob and I could see the instrument panel from where we were sitting, and once we were airborne, he leaned over and said to me, "Look at the air speed indicator."

I peeked over the pilot's shoulder and saw the airspeed indicator was over the red line; the pilot was trying to make up the time we had lost because of his failure to release the parking brake, and he was pushing the plane beyond its normal limits.

"What in the devil is this guy thinking?" I said to myself.

As soon as we got on the ground in Wichita, I went to the flight standards office to file a complaint. I knew an inspector in that office from my days in Wichita. He wasn't in just then, but the office chief came out and asked, "What's the problem?"

I told him the story of what had happened in Kansas City, and he looked at me and snarled, "You guys are nothing but trouble-makers."

I responded, "Oh, is that what we are? I thought we were safety inspectors."

"No," he retorted, "you're troublemakers."

"You can think what you want," I answered, "but when I get back to Kansas City, I'm going into the Flight Standards division chief's office and report this to him."

"Why would you do that?"

"Because your pilots are violating basic safety rules."

So, I filed the report. I'm not sure how far it went or if anything ever happened to that pilot, but I did what I felt was the right thing to do at that particular time. That was just how I dealt with everything in my flying career.

And that brings me to one of the stranger projects on which I worked while in the Atlanta ACO. In late spring of 1992, I was sent to China to test an airplane under production there. The U.S. Department of State sanctioned and encouraged my trip as part of an effort to mend U.S.-China relations after several years of economic sanctions imposed in the wake of the 1989 Tiananmen Square Massacre.

The State Department arranged with the FAA to send over

teams from flight test, systems, propulsion, and maintenance. I was told, "We'd like you to go over there and give us your opinion on whether the aircraft is certifiable." The clear implication was that they wanted me to come back with the certification.

I told my boss I would go to China but would not certify the airplane unless I truly believed the airplane warranted it. I don't know why I always got these strange projects; maybe this time I had volunteered.

Once I got to China, I spent several weeks in Beijing giving ground instruction. However, the manufacturing plant was in Harbin in the northeast corner of the country. Rich Adler, a flight engineer from the Chicago ACO, went with me to do all the technical calculations on the aircraft. The day we were scheduled to fly to Harbin, a man from the U.S. Embassy in Beijing came to our hotel and took me aside.

"Dave," he said, "when you're up there in Harbin, we'd like you to come back and tell us what you saw."

I knew exactly what he meant, but I played dumb and responded, "What do you mean 'what I saw'?"

"Tell us about their nav aids and about this and about that." I knew the "this" and the "that" was anything related to the Chinese military or the country's national security apparatus.

"I'm not going to do that," I stated firmly.

"Why not?"

"Because that's what spies do, and I'm not a spy and I don't intend to be one."

"Oh, we don't want you to spy," he countered with a sly grin.

"OK, but I'm still not going to do it."

It did not matter what he called it; it was still spying, and I was not about to go along with it. The guy was decent enough to drop the conversation there and did not appeal to my sense of patriotism. Even if he had, I would not have given in because that was not my job. Even though I was working for a federal agency, if I had given in to his request, there is no telling how much more he might have wanted me to do. Besides, I did not want to make my ex-wife's suspicions about my being a spy become a reality.

Harbin is not exactly a tourist mecca. It has an annual ice sculpture festival but other than that, there is not much there to see or do, at least when I was there. It now has a high-rise 5-star J.W. Marriott Hotel and a Holiday Inn, but in 1992 the supposed 4-star hotel they put me in was more like a one-star or a no-star back in the States. They originally tried to make Rich and me sleep in the aircraft manufacturer's barracks with the workers, but I was having none of that.

We had a driver who would take us back and forth to the plant. One morning I was in my room waiting for the driver to arrive, and the housekeeper came in. I had taken a shower and left my towel on the floor. She came in, picked up the towel and went to work. I observed her as she first wiped off the countertop around the sink. Then, she wiped off the top of the toilet seat. Then, she wiped out the toilet bowl. Finally, she took that same towel, wiped out my drinking glass, put it back on the counter, and considered it clean. I put that glass aside and never used it again. After that, I made sure I personally washed everything before I used it.

Back then, I wore white socks with my flight boots, and during the three weeks I was in Harbin, I could not figure out why the soles of my socks became blacker and blacker. The room was carpeted, but eventually I discovered when the housekeeper cleaned the carpet, she didn't use a vacuum; she used a mop. She just kept spreading the dirt around, and I was walking around the room in my stocking feet gradually absorbing that dirt.

The airplane we were sent to test was the prototype of what

would become the Harbin Y-12, manufactured by the Harbin Aircraft Company. It was a twin-engine high-wing turboprop that seated 12, plus crew. It was much like a small commuter plane.

That airplane's configuration was unbelievable. Virtually all other aircraft use hydraulic systems to raise and lower the flaps and operate other systems. This airplane was using bicycle chains to raise and lower the flaps. It was like something from the 1930s—not a very sophisticated airplane. It did not look good and felt worse when it was in the air. It was not an enjoyable plane to fly. It was a disaster waiting to happen.

The chief pilot for Harbin Aircraft could not speak English and I could not speak Chinese, so we had an interpreter fly with us every day. The interpreter sat in the jump seat between the Chinese pilot and me. What I quickly discovered is that trying to fly an airplane through an interpreter does not work very well, especially when the interpreter is not an aviator.

One morning we were up flying and suddenly I heard the interpreter throwing up. He was spilling his guts into a barf bag. The Chinese pilot just shrugged his shoulders and I thought sardonically, "This is going to be a lot of fun."

On another flight, I could see the Chinese pilot getting upset and starting to sweat. I asked the interpreter what was wrong.

"We are lost," he answered.

At that time, the navigation aids in that part of China were nothing like we have here in the States and in many other parts of the world. Most modern navigation aids are equipped with an Automatic Direction Finder (ADF), which allows the aircraft to home in on a particular station. There was nothing like that in China in 1992. To say the Chinese navigation aids were primitive is giving them more credit than they deserve.

Even without proper navigation aids, I had a good sense of

what our route had been and where we were. As the pilot continued to sweat profusely, the more nervous the interpreter became. Finally, I looked around, tapped the pilot on the shoulder, pointed out the window and said, "Airfield right there." He let out a big sigh and wiped his brow. I don't think the interpreter relaxed until we got on the ground.

One day we were testing whether the plane could get into the air with one engine. The Chinese pilot was doing the flying and as he was going down the runway, he decided he was not going to make it on one engine and decided to stop. Well, there was no way he was going to stop on that runway at the speed we were going, so I grabbed both throttles, slammed them to the firewall and yelled, "I got it!" I took over the airplane and got us off the ground. I don't know what he was thinking, but he obviously had no confidence in the airplane or in his own skills as a pilot.

One of the major problems with that airplane was that Rich Adler was never able to get what pilots refer to as a V speed, which is the speed needed to get the plane into the air—takeoff. When you determine a V speed for an aircraft, you're supposed to hit that number every time you fly that plane. That *particular* airplane, though, before it ever got to the speed at which you can no longer abort the takeoff wanted to jump into the air. I had to hold it on the ground so I could get the necessary speed to take off.

The vice president of engineering for Harbin Aircraft was extremely curious about our test flights. After each one, he would ask me, "Well, how did it go?" I usually had to tell him we had one problem or another and he'd exclaim, "Oh, no!" and smack himself on the side of the head. I am not sure if he knew that plane had problems or was hoping we would not discover them.

The day we left Harbin, the vice president of the company told me he wanted to hire me to come back and serve as a consultant.

I told him "No" in no uncertain terms. The place was primitive. It was like living in the 1900s, even though the city had more than a million people. There were fields around Harbin that looked like Kansas or Nebraska, but when workers tilled the fields, they didn't do it with tractors; there were people using shovels and hoes. When you're tilling millions of acres with shovels and hoes, you're a little bit behind the times.

After four weeks in Beijing and three weeks in Harbin, Rich and I were ready to go home. When I left Harbin, I said to myself, "If I never see this place again, it will be too soon." Not only was the place primitive, but they also had some of the most godawful-tasting booze, no matter where you went in the country. When we went out to dinner in Beijing with our hosts, there usually would be six or seven Chinese and the two Yankees. Each of the Chinese would propose a toast, which meant another round of that awful booze. After just a couple of rounds, Rich would say to me, "Dave, you can do the toasts from now on because I can't take any more of that stuff."

Not long after I got home someone from the FAA called me and said the agency wanted me to go back to Harbin with the certification team. My answer was short and sweet: "No."

"Why not?" I was asked.

"If I go back, the airplane will never get certified. I won't certify that airplane. There are too many things wrong with it."

"The State Department wants it."

"I don't care what the State Department wants," I said. "I'm not going to certify it."

So, the FAA picked this young kid from the Los Angeles office who had absolutely no experience. They sent him over there, and the airplane got certified. I don't know how they ever got the numbers to work out. And I probably don't want to know.

I made several trips to China for various projects while I was working in the Atlanta office. One was to Shanghai in eastern China, one to Guangxi in the south, and one to the western part of the country to a company that flew DC-9s and was making modifications to them. That company had a full-time Boeing rep who lived there. I have no idea how he survived because western China was even more primitive than Harbin. The Chinese would have had to pay me a bundle of money to work in that place.

Not all my foreign trips while I was with the FAA were as strange as those to China. In fact, my trips to Brazil to work on projects for Embraer were among some of the better assignments, primarily because Embraer was much more professional about its business and its aircraft. I started going down there in 1982 and continued until the company, which originally was government-owned, was sold to private investors in 1994.

Embraer workers were thrilled when that sale took place because finally the company had some solid financial backing. Through the 1980s and into the early 1990s, inflation in Brazil was out of control, roughly 10 per cent a month. From one trip to the next I never knew how bad things were going to be.

In 1994, the same year Embraer was sold, the country changed its national currency from the *cruzeiro real* to the *real*. I was in Sao Paulo the day they started delivering the new currency; every bank had armed guards in front of it. Since that sale, Embraer has gone on to become the third largest manufacturer of civilian aircraft in the world, behind only Boeing and Airbus.

During those years when the Brazilian government owned Embraer, and with skyrocketing inflation, the company struggled financially. When it sent representatives to meet with FAA officials in Atlanta, they had to stay two to a room to save money. But the Embraer employees loved the trips to the States because they would load up on things they could not buy in Brazil due to inflation: VCRs, disposable diapers, and anything else that was in short supply and high demand back in Brazil.

One weekend Becky and I decided to invite the Embraer officials and some local ACO co-workers to our house for a cocktail party. We had appetizers spread out over the dining room table and a fully stocked bar in the kitchen. By the time the party was over, every bit of food and every drop of liquor was gone. The Brazilians just ate and drank continuously; it was as if they had not eaten for several days. We had virtually no food left in the house: there was not a saltine cracker, pickle, or olive to be found.

My work with Embraer primarily was as the project pilot for the EMB 120 Brasilia, a twin-engine turboprop commuter aircraft popular with a number of airlines in the U.S. While my relationship with Embraer officials was always professional, just getting to the work site could sometimes be a bit of a challenge, especially if I tried to get there on my own.

Embraer has its headquarters in the southeastern part of the country about 60 miles northeast of São Paulo. On one of my later trips, I decided to take Becky along. We stayed at Ipanema Beach in Rio de Janeiro for a few days before renting a car to make the roughly 220-mile drive to São José dos Campos.

The guy at the car rental place did not speak any English, and neither Becky nor I spoke any Portuguese, so our efforts to determine how to get on the road to São José dos Campos were lost in the lack of translation. (Remember, I took Portuguese at the Naval Academy, but I had forgotten most of it by 1994.) Finally, we were able to figure out how to head in the general direction of the highway they called the Rodovia Presidente Dutra to São José dos Campos.

Before we even got on the highway, though, we encountered a

traffic circle, or roundabout, with no idea where to get off and get on the road to the Embraer plant. Finally, we saw a sign that said "Ford" and thought there might be somebody there who spoke enough English to help us. So, we went into this Ford dealership, and Becky with her limited Portuguese asked one of the sales staff if he spoke English. He did not.

However, he knew how to speak French. Becky also spoke French and was able to elicit enough information to get us onto the Rodovia Presidente Dutra on our way to São José dos Campos. The highway, as it turns out, was not much of a highway. It was more like a country road with a fruit stand every half mile. Eventually, we found São José dos Campos and the Embraer plant, and I was able to complete the project I was sent there to do.

By the time Embraer was sold to private investors in 1994 and my work there ended, I was beginning to get restless and was tiring of the bureaucracy. I wanted some new challenges. I already had my 20 years in with the FAA, including the year I spent in the Cerritos, California, office before going to Northrop. I was looking for an opportunity that would be a bit more challenging than what I was doing then. It was not so much a matter of where I would go after I left the FAA—but when.

CHAPTER 12
THE PERFECT UMBRELLA BOSS

There is a tendency among many Americans to look at the federal government as something akin to a benevolent dictator whose primary function—including the imposition of income taxes—is to ensure the health and well-being of his citizens. Anyone who has spent any time working for a federal government bureaucracy knows such is not the case. The bureaucrats who run the bureaucracies probably make more decisions in favor of businesses with which they interact every day than what is in the best interests of the average citizen.

That was especially true with the FAA when I worked there, as exemplified by my former boss, John, who favored Gulfstream interests whenever there was a dispute involving the company. Because of his favoritism toward Gulfstream, the company eventually rewarded him with a high-paying job. John would have done anything Gulfstream asked him to do, which is why we got into that major confrontation I wrote about in a previous chapter.

Bureaucracies have devolved to the point where they tend to look out for their own interests and the businesses they are mandated to oversee rather than being watchdogs for the interests of the average citizen. This behavior was seen frequently when the FAA dealt with the big manufacturers such as Boeing, Piper, Cessna, Gulfstream, or McDonnell-Douglas; they received favored treatment while the small companies got hammered.

Over the years, the FAA convinced Congress it did not have the manpower to do all the inspections necessary on new aircraft and modifications to existing aircraft, especially with the major manufacturers. The manufacturers said the only way the work was going to get done was to have Designated Engineering Representatives (DERs), working for and paid by the manufacturers to do that work, supposedly with oversight from the FAA.

To think a DER would override a decision his company makes to save time and money is being incredibly naïve. If you work for a company and it gives you a check every two weeks, who are you going to be beholden to: the FAA or the company paying you? It does not take a genius to figure that one out.

A similar situation exists with Congress. Boeing and the other major aircraft companies donate a lot of money to Congressional campaigns, and elected officials are not going to crap in their own mess kits if there is a dispute between the FAA and the manufacturers.

Bureaucracies also have a tendency to address small problems that affect just a few people with grandiose programs that throw everyone under the same cloud of suspicion. Such was the case in the early 1990s when the FAA introduced mandatory drug testing because of concerns that a spike in illegal drug use among the civilian population would be reflected among federal workers, especially pilots.

When you have a social problem, where do you start to try to solve it? You start with the military and government workers because they are a captive audience who cannot refuse mandated requirements such as drug tests. Under what was known as the Omnibus Transportation Testing Act of 1991, the FAA mandated monthly drug tests for randomly selected pilots.

Some of this may have been prompted by a 1990 incident in which the pilot, co-pilot, and flight engineer of a Northwest Airlines 727 were convicted of flying while drunk after a flight from Fargo, North Dakota, to Minneapolis, Minnesota. The three had their blood-alcohol levels tested two hours after landing following a complaint that the three had been seen drinking heavily the night before the flight. All three registered above the 0.04 percent level permitted by the FAA.

The joke that circulated among pilots after that incident was, "How many pilots does it take to fly a Northwest 727?" The answer: "Two and a fifth."

The drug tests in the Atlanta office turned out to be not-so-random because every month my name was on the list of pilots to be tested.

"How in the devil can that be random?" I asked incredulously.

"Well, that's just the way it goes," I was told.

The first time I went in for a test in the Atlanta ACO, the female staffer at the desk asked me for my identification. I looked at her and said, "I'll show you my ID when you show me your ID. And show me a piece of paper that says you are authorized to take my specimen."

"I'll make a call and get that authorization," she said.

I sat around for several hours waiting for that authorization. Finally, about two o'clock in the afternoon, a member of the testing company came in and showed me the paperwork and said, "Now you need to give us a sample."

"I don't have to go to the bathroom now," I answered with a smile.

"Well, go drink some water." They wanted me to drink eight ounces of water every 30 minutes. So, I drank and drank and still did not go to the bathroom. Finally, about 4:30, the drug testing staff packed up and left.

Paul Sconyers, the assistant office chief, was with me and said, "Dave, they just left. You can go to the bathroom now."

So, I did...for a long time.

The next time I was called in for a specimen, the staffer had her ID and a letter authorizing the testing company to take my specimen. But I told them, "For gosh sakes, you can't be too much more conservative than a test pilot. We're not really on drugs."

"Well," the staffer said, "we've got to be sure."

"You're barking up the wrong tree, lady," I added.

About that same time, the airport began using drug-sniffing dogs to try to find illegal drugs on passengers and aircraft. One day I was observing a test with a drug-sniffing dog to see if it could pick out a planted package of drugs. Passengers were getting onto the plane and the dog was going up and down the aisle sniffing bags and people, looking for that hidden package. Just when the dog got to the galley area on the 727, he stopped and took a dump in the middle of the aisle.

As soon as the dog made his deposit, he jumped over three rows of seats and found the package. The captain, who was also watching the test, put his hand to his head and said, "Aw shit." It would be a while before that plane took off because a cleanup crew had to be brought in.

In December 1985, Atlanta hosted a conference for DERs from throughout the Southeast. The idea was to bring them in, go over a bunch of rules changes, create some camaraderie by getting to know one another, and have a few drinks.

As part of the entertainment for the conference, the Flight Test branch decided to have a performance of the "Navel Quartet." I am not sure if it was Bob Sample or I who came up with the idea, but we left the impression with the DERs planning to attend that it would be a classy act, perhaps something from the Naval Academy or Navy headquarters in Washington. Instead, the quartet came from the "*Navel* Academy."

Bob, I, and two others wrapped leather flight jackets around our waists, took off our shirts, and painted noses and eyes on our chests and bright red lips around our navels. Then, we fashioned top hats out of black plastic garbage bags and put them over our heads to hide our faces. We did a little dance to the "Colonel Bogey March" (made famous as the theme song from the movie *The Bridge on the River Kwai)* and left the crowd in stitches. It was not particularly classy, but it was a lot of fun, and it was just the kind of thing pilots do to entertain each other. I even have a photo to commemorate the event.

I was in charge of the hospitality room for that conference and planned to stay there that night after having a bit too much to drink rather than making the long drive home. One of my pilots was a woman, who some weeks earlier had come to me with a complaint. "Dave, I'm having problems with the other pilots."

"Why? What's going on?" I asked.

"We go out for drinks after work and all they want to do is tell raunchy stories." She explained she was offended by their rough language.

I told her she did not have to go with them, but she countered, "But, I learn a lot by listening to their stories that involve situations they've encountered while flying."

"Well, then just shut up and listen. You can't have it both ways."

That night of the conference, this woman also had had too

much to drink and said she wanted to stay overnight in the hospitality room. She lived in Cobb County on the other side of Atlanta and, like me, thought better of driving home.

"OK," I said, "but I'm staying here, too, and there's only one bed. So, when we get into bed, you stay on your half, and I'll stay on my half."

She looked at me for a minute, thought about it, and said, "That's OK. I think I'll go home with John Henderson and his wife."

The woman eventually left the FAA and went to work for Piedmont Airlines as a pilot.

Maybe it was because our profession was so inherently dangerous that when we let off steam, we really let off steam as we did at the DER conference. But the idea that something could go terribly wrong at any time we were testing a new aircraft or an aircraft modification was never far from the surface. What made that idea even more difficult to absorb was that when something did happen, the FAA or the aircraft manufacturer could be incredibly callous.

One of the stories I heard in the late 1970s involved a company pilot by the name of Bob Stone. Bob was the favorite pilot of Olive Anne Beech, the wife and co-founder of Beechcraft with her husband, the famed Walter Beech. Whenever Mrs. Beech wanted to go somewhere, she would take Bob out of the test community and make him her private pilot. Bob probably did a good share of Beechcraft's critical testing, except when Mrs. Beech wanted to fly somewhere.

As the story goes, Bob was on a test flight with a T-34C Navy trainer in the late 1970s when something went wrong, and the airplane executed a violent nose-down pitch with negative G-forces. Someone estimated it might have been as much as a minus 10 Gs, which would have been enough to break his neck and kill him instantly. I am not sure if that is the case because I have never seen the accident report.

However, I was told the company refused to pay Bob's family his hazardous duty pay. Their rationale was that he did not qualify because he did not finish the test. Bob had been loyal to the company, but that loyalty was not reciprocated.

The FAA could be just as callous. Several years after I got to the Atlanta office, one of our pilots, John Henderson, was killed in a crash during a flight test in Alabama.

John was an excellent pilot and extremely thorough about his testing. He was not a loosey-goosey pilot who was prone to make mistakes. But this particular test was a spin test on a single-engine plane to see what a pilot would need to do to recover from the spin. When he started the spin, the ballast in the aircraft shifted because it was not properly secured. John could not recover the aircraft and decided to bail out. Something went wrong and he did not survive.

When word reached me about his death, I immediately called the FAA human resources office in Kansas City and told them we needed a casualty assistance team to fly to Atlanta the next day; John's widow and his family needed guidance on what assistance the government would provide, what the benefits would be, and what the government would and would not do to help see the family through the tragedy.

Having come from the Navy, I was familiar with the term Casualty Assistance Officers and what their duties were. In the military, the response to deaths is almost universally prompt and respectful. The response I got from a young woman in the Kansas City office was so cavalier I was infuriated because it was disrespectful to John and his family.

She told me there was no way they could get to Atlanta the next day and it might be another week before they could get anyone there to help the family.

"Well, sweetie," I answered, "I know when the first airplane from Kansas City arrives in Atlanta tomorrow. If you're not on it, I am going to personally call the director of the FAA's Central Region and ask him what the hell is going on that you can't come to assist a family that just experienced a death."

She hemmed and hawed a few minutes and finally said, "Let me get back to you."

A little while later she called me back and said, "We'll be on that flight."

"I thought you would see it that way," was my response.

Eventually, everything was taken care of. After the funeral, John's widow came up to me and said, "John told me, 'If anything ever happens to me, talk to Dave. He's tough, but he's fair.'"

I was especially tough on people who worked for me when they demonstrated they were not always trustworthy or capable of doing their jobs as I felt they should be done.

One of these individuals was an aerospace engineer who shall remain nameless for the purpose of this book. He was a GS-5 who flunked out of air traffic control school. Someone from ATC came to the FAA certification office and said they would give us this guy and would not charge him to our personnel total if we would just take him off their hands. He was a disaster. You never knew what he was doing or where he was going, but he fit in well with the bureaucratic standard of performance, or under-performance, within the FAA.

Another poor soul in over his head was a mechanical systems certification engineer who also shall remain nameless. He was very unreliable. If he had an appointment with an applicant at 9:00 a.m., the applicant would be there, but in many instances the engineer would not. In addition, nobody knew where he was at any particular moment. Usually, later in the day he would provide some sort of explanation about where he had been.

On one of those occasions when he had missed an appointment with an applicant, he told me he had gone to Florida to visit his mother because she was not feeling well.

"Did you bother to tell the people you were supposed to meet that you wouldn't be here?" I inquired.

"No."

"Why not?"

"I didn't think of it."

One day he came to me about an opening we had for a GS-14 flight test engineer. He wanted to be considered for the job. I reminded him of all the previous conversations we had had about his being unreliable.

"Well, I'll change," he pleaded.

I told him that if he changed, I would consider him for the next job but not for this one.

"You're not ready for it," I said, ending the conversation.

During my tenure as chief of the Flight Test branch, there was never a dull moment. Handing out travel assignments, reviewing individual reports, and dealing with a variety of personnel issues could sometimes be stressful, so a bit of humor was always welcome. Bob Sample, one of the pilots, kept us entertained with what he referred to as "The Who G-A-S Log," the "G-A-S" being an abbreviation for "Gives-A-Shit."

An engineer might come into the office one day and announce, "Over the weekend I had to buy a new set of tires for my car, and they set me back $300." Bob would invariably reply, "Well, that's one for the Who G-A-S Log." Or one of the pilots would remark at a meeting that "Before I came to the FAA, I had flown more than 500 hours in twin-engines." Bob would pipe up, "That's one for the Who G-A-S Log."

After Bob died in 2007, his wife, Joan, asked me to deliver the eulogy at his funeral. Although I've never been afraid to speak my mind at a meeting, getting up in front of an audience and speaking is something that always terrified me. Nevertheless, I agreed to do it and decided to talk about Bob's "Who G-A-S Log." Bob was Catholic and a priest was officiating at the service so I turned to the priest and said, "Sorry, Father, but I need to mention the Who G-A-S Log and I bet even you can figure out the title with a little help: Who ... gives ... a ..." The priest smiled, turned, red, and joined everyone else in laughter as I told the story.

By 1996, I knew it was time for me to leave the FAA and move on to something where I could have more control over what I did and when I did it. I had 37 years of government service and was getting up in years. I did not want to hang around the office until I was 90 years old. Besides, I knew I could make more money as a private consultant than I was making in the FAA and was sorely in need of a change of attitude and atmosphere from what I was experiencing inside the bureaucracy.

I put in my papers and left the Atlanta FAA office in January1997. At my retirement party one of my pilots, Edgar Wilson, paid me a great compliment when he described me as "the perfect umbrella boss."

"Why do you call me that?" I asked.

"Because you are holding the opened umbrella above us and under the elephant's tail, so when the elephant tries to shit on us, it just rolls off."

It was a nice way to leave government service and head to the private sector, where I learned that umbrella would come in handy on a number of projects over the next few years.

CHAPTER 13
THE PHOENIX FANJET FLAMEOUT

Within a month of retiring from the FAA in January 1997, I was back at work in the aviation industry, this time as a Designated Engineering Representative. There are two types of DERs: company DERs and consultant DERs. A company DER works for a specific aviation company and, according to FAA regulations, can only approve, or recommend approval of, technical data on behalf of the company. Consultant DERs are independent contractors hired by companies for specific aircraft-related projects. Consultant DERs run tests and then make recommendations to the FAA about whether it should approve or reject specific projects based on the technical data the DERs have gathered.

I chose the consultant DER route because it gave me more freedom to choose which jobs I took and which ones I passed on. I figured going the consultant route would enable me to take on a wider variety of jobs of my choosing and set my own rates for jobs rather than tying myself to a single company and its pay scale. Being a consultant DER was far more lucrative than being a company DER. I could work when and where I wanted at the rates I wanted to charge.

The DER ranks are split into nine technical disciplines: acoustical engineering, engine engineering, flight analyst, flight test pilot, power plant engineering, propeller engineering, radio engineering, structural engineering, and systems and equipment engineering. I applied for and was certified as a flight test pilot DER.

My first job as a consultant DER was with the Canadian-based company Alberta Aerospace, which was trying to develop a small jet trainer it had dubbed the Phoenix Fanjet. The operations manager of the project was Ray Dickey, a former engineer at Piper Aircraft in Vero Beach, Florida. I knew Ray fairly well, and he had given me a heads up on this job opening. When I interviewed for the job with company's chief executive officer, John McIntee, he looked at my résumé, let out a low whistle, and said, "I would *kill* for your credentials."

McIntee's intent was to get the Phoenix Fanjet certified by the FAA as an *ab initio* trainer, meaning it would take a pilot from zero hours of flight time to the point he or she qualified for advanced jet training. Most civilian flight training schools used a combination of single- and twin-engine piston aircraft or an occasional twin-engine turboprop. That process was time-consuming; it could last as much as two years before pilots moved on to advanced jet training. The Fanjet was designed to cut training time and costs.

The Fanjet had another purpose, though, which was one of the features that attracted me to the job. Many pilots never get to see the bad aspects of an aircraft's performance, known as "upsets." They are never really taught how to get out of tough situations because putting them into those situations in training is considered too dangerous unless they are conducted in the safety of a simulator.

In other words, pilots rarely get actual in-air flying experience in figuring out how to get an airplane out of trouble, whether it is a stall, a spin, or issues on takeoff or landing. So, when pilots experience some type of problem, it is most likely the first time it has happened to them, and usually they do not know how to handle the upset. For many pilots, as long as they hit the V speeds and are able to fly at 35,000 feet, they are happy. If something unexpected happens, they don't know what to do unless they've been through it a number of times in a simulator. The process is known as Upset Prevention and Recovery Training (UPRT).

Any airplane used for upset training would have to be quite agile and incredibly forgiving. McIntee believed the Phoenix Fanjet could be that type of aircraft. He was getting interest from Aeroméxico, Delta Airlines, Northwest Airlines, and Embry-Riddle Aeronautical University in Daytona Beach, Florida, about the Fanjet's potential as a trainer for fledgling pilots.

McIntee and Ray Johnson, a former Piper executive who had signed on with Alberta Aerospace as the company's president, believed that once the FAA certified the plane, they could produce and market them for a little more than $1.5 million each. Plans were to manufacture up to 150 planes every year at a plant they intended to build in Claresholm, a town of fewer than 5,000 people about 80 miles south of Calgary in western Canada.

The Phoenix Fanjet was the latest iteration in a series of experimental single-engine jet trainers that had come out of the workshop of noted Italian aircraft designer Stelio Frati in the 1950s and 1960s.

The first of Frati's experimental aircraft was the Caproni Trento F.5, a wooden, two-seat turbojet with a top speed of less than 200 knots. The second was the F.400 Cobra, another wooden two-seater with a top speed of 310 knots. Next came the Jet Squalus F.1300, designed by Frati for the Belgian company Promavia, which wanted to sell it to the Belgian Air Force as an *ab initio* trainer; however, the Belgian Air Force gave up on it after only six test flights.

The Squalus also was designed to appeal to the U.S. Air Force as its Next Generation Trainer. It featured the side-by-side seating configuration the Air Force preferred in its trainers and had been using that configuration in its T-37 Tweet primary jet trainer for years. That plan fell through when the Navy and the Air Force in the late 1980s agreed to cooperate on development of the Joint Primary Aircraft Training System (JPATS), which featured the more traditional tandem seating arrangement, which the Navy preferred.

Promavia did not give up trying to market the Squalus after failing to entice either the U.S. or Belgian air forces, though. In 1993, it established a partnership with the Mikoyan Experimental Design Bureau out of the Soviet Union and Boeing out of the U.S. to come up with a derivative of the Squalus. That project was known as the ATTA 3000. However, the ATTA 3000 never got off the ground—literally—and Promavia went out of business. It was then that Alberta Aerospace, one of Promavia's creditors, went to court to gain access to the rights to the plane. A protracted legal battle eventually stretched across two continents and several countries before the courts decided in favor of Alberta Aerospace.

By the time I got involved in the Fanjet project in early 1997, the lone prototype of the plane had been shipped, not flown, from Europe to Canada. During its unloading from the cargo ship, the plane apparently was dropped, causing a great deal of damage that had to be repaired before it was sent to Calgary.

The first time I flew the plane in mid-January 1997 was the first time it had been in the air since it left Europe. Since the Fanjet had recently been damaged and was still considered experimental, I knew I had to be extremely careful just getting it off the ground.

In a situation such as that, you don't just jump into the cockpit, fire up the engine, and take off. You have to ease into it and take it slowly to make sure the plane is actually capable of doing what it is supposed to do.

The first thing I did was start the engine and look at all the instrumentation to check the readings. Next, I taxied slowly around the airport. Then, I did some high-speed taxis. I ran it up to 60 knots, and then to 80 knots, and finally to 100 knots until I felt comfortable the airplane was going to fly.

Even before I got the plane into the air, though, I realized it had at least one major problem that would take time and probably a great deal of money to fix. Since I am not an aerodynamicist, I am not sure exactly why this was happening, but I discovered if I over-rotated slightly on takeoff (lifted the nose too quickly), the airplane staggered and wanted to mush back into the ground. I had to ease it off the runway; I could not just yank it into the air as I could most airplanes. That problem needed to be corrected before the company could even begin to think about getting the plane certified as an *ab initio* trainer.

The weather in Calgary during the winter is not particularly conducive to testing aircraft. In fact, winter weather in that part of Canada can be snowier and colder than where I grew up on Michigan's Upper Peninsula. So, a decision eventually was made, partly at the urging of Ray Dickey and Ray Johnson, to do the testing where it was warmer and our testing days would not be limited by snow or cold. Since both the Rays had lived and worked in Florida, they decided we should do the testing there, specifically in Lakeland, just west of Orlando.

Getting that airplane out of Canada was a great relief to the Canadian government, for reasons I never fully understood, unless it was just tired of seeing that plane. I only flew the Fanjet three or four times in Canada but was told rather bluntly by one Canadian government official that, "The next time you take off, we want you going across the border to the United States."

On January 29, 1997, with the temperature 30 degrees—relatively balmy for that time of year—and wind gusts up to 25 knots, I got into the only existing model of the Phoenix Fanjet for the trip to Lakeland. It was a nice-looking plane—white with a horizontal red stripe on the fuselage from nose to tail and a large red Canadian maple leaf on each side of the tail. McIntee decided to go with me, flying in the left seat. I guess he just wanted to make sure his investment got to Florida.

Before we left Canada, McIntee decided to buzz the small airfield at Claresholm. I thought that was lousy idea since the Canadian government already had a poor opinion of the plane, and I let him know.

"John," I said as diplomatically as possible, "you don't want to be buzzing this field unless you want to get violated by the Canadians."

"Oh, they all know me here," he said dismissively.

"But they don't know me," I responded, not so diplomatically. "You better knock this off, and we'd better go about our business."

McIntee did not care what the Canadian government thought about him or the Fanjet. Nor did he think much of my advice. He wanted to impress all his buddies and anyone who might want to invest in the plane and the manufacturing plant he had promised to build in Claresholm. This was a promotional stunt for him because he was in desperate need of investors, as I was to learn later.

Someone, John I suspect, made a call before we left Calgary and let another someone know that despite my warning, he planned to buzz the airfield at Claresholm. When we got there, I saw about 50 people standing around in the cold and wind, looking at the plane.

John got on the radio and apologized for not being able to make a brief stopover because of the high winds.

"Thanks for coming out, folks," John said. "We'll bring you back some pictures when we're down (in Florida). Take care."

He buzzed the field three times and then we headed southeast for Minot, North Dakota, where we refueled before flying on to Lincoln, Nebraska, to spend the night. The next day we headed southeast again before landing at the small airport in Lakeland, Florida, in absolutely beautiful weather.

The first order of business after we got to Florida was to change the plane's engine. The original engine was a Garrett TFE109 turbofan, which had been used on only one other aircraft, the American Fairchild T-46 trainer. The T-46, which was designed to be the Air Force's Next Generation Trainer in the 1980s, never got much past the drawing board and only three models were produced, partly because JPATS rendered it obsolete.

Replacing the Garrett engine was the Williams FJ-44, which was smaller and lighter but had more thrust and was much more widely used in a number of aircraft models in the U.S. Despite the FJ-44's better record, I was not sure how it would work on the Fanjet, given the fact the plane already had a propensity for balky handling on takeoff.

Before I flew the Fanjet for the first time with the new engine, I called my son, David, and asked him to be there that day in case anything happened to me. Becky was still back in Georgia teaching school, but I wanted a family member on the ground in the event the airplane had issues I could not resolve. I just wanted him there on site; I was not about to put him in that airplane.

The problem on takeoff did not go away with the new engine. I certainly did not have a clue about what was causing it to stagger, although it may have had something to do with the angle of attack.

One day I had an aerodynamicist fly with me to see if he could get a fix on the problem. Before we took off, I explained in some detail what the plane was doing.

"Can you demonstrate to me what you did to make that happen?" he asked.

I told him I could, but that we were going to use a long runway because if it started staggering again, I was going to set it down and we were going to stop.

It happened again and he looked bewildered. "That's the strangest thing I've ever seen," he commented. He said there was something wrong somewhere in the airplane, but I never found out what it was and I'm not sure he did either.

At some point during our months of testing, Ray Dickey came to me and said he wanted the chief pilot from a local flight school to fly with me. I told Ray I did not think that was a good idea. When he asked me why I replied as patiently as I could, "This is an experimental airplane, and we're having some problems with it."

"Aw, he'll be OK."

He was not OK. He was no more than 30 years old and had little flying experience. As we started down the runway the plane was staggering, just hanging on.

"I'm going to raise the flaps," he said.

"If you raise those flaps, I'm going to break your goddamn arm!" I shouted. "We're going to fly out of this thing. If you raise the flaps, it's going to be even worse because the stall speed will go up. It won't get better." Here's the explanation for non-pilots: raising the flaps on takeoff decreases lift and increases the stall speed. But this guy apparently thought raising the flaps would get rid of the stagger.

We barely cleared the hangar, and the control tower came up

on the radio wanting to know why we were hot-dogging it. "Hot dog, shit!" I yelled back at them. "We've got a problem!"

We flew for a while and then landed without incident. That kid never came back for another ride with me.

I rented a small apartment in Lakeland to be close to the airfield because I was flying the Fanjet on a regular basis, sometimes twice a day when it wasn't down for maintenance. About two months into the program, I called Becky and convinced her to quit her teaching job 12 weeks before the end of the school year and join me in Lakeland, which she did.

Since the Phoenix Fanjet was a one-of-a-kind plane, with no backup, when maintenance was needed, I had some down time. On those occasions, I was able to take short-term consulting jobs with other companies because I had that specific stipulation written into my contract with Alberta Aerospace. (I will go into some of the more interesting consulting jobs I had during the 18 months I worked on the Phoenix Fanjet project in more detail in the next chapter.)

While the testing was going on, John McIntee was pushing hard to lure investors and get the airlines interested in the potential of the Phoenix Fanjet. Chief pilots and chief instructors for various airlines showed up in Lakeland to check out the plane on a regular basis even though it was a long way from meeting Federal Aviation Regulations and being certified.

McIntee even brought in some aviation journalists, some of whom were pilots, for test rides to help spread the word about the Fanjet's potential. Those journalists usually were rather complimentary of the plane, although they did tend to point out some of the difficulties we were encountering. One writer referred to me as being "professional but effusive." I don't think anyone other than this guy has ever referred to me as being "effusive." If anything, I am the antithesis of effusive.

We also brought in a lot of people to test-fly the plane who should not have been there because they just did not have the skills necessary to fly an experimental aircraft. I always wanted these pilots to see the airplane stall. The Phoenix Fanjet was a reasonable staller and could be recovered easily.

A lot of these guys would get within 10 miles an hour of the stall speed and say, "There it is."

I would laugh and say, "You ain't even close."

While the idea behind the Phoenix Fanjet was admirable, it eventually became apparent that the work necessary to get the plane to the point that it could do everything McIntee wanted it to do was far beyond the company's technical abilities and financial resources. McIntee was in a hurry to get the plane certified and into production but only saw the result, not the steps and money it would take to get there.

What I did not know when I took the job was just how under-financed this operation was. The company had managed to put together only $10 million in investments to get the airplane certified. When I learned this, I told them that amount of money was peanuts; they were never going to get the airplane certified with that meager amount. Nothing is cheap when it comes to testing an airplane, especially when you're trying to work out the bugs in order to get it certified.

Company officials said they already had all the data they needed, which would cut costs. When I asked where the data was, Ray Dickey said they had sent a team to Italy to get it from Stelio Frati. The only problem was all the data was in Italian, and Alberta Aerospace did not have anybody to translate it. It probably would have taken a million dollars just to complete the translation of all the technical data.

By the summer of 1998, it was clear there was not enough

money to fix all the aircraft's problems. The company just ran out of money. I figured the end was near when they told me they were going to ferry the plane back to Canada. My contract was up in August 1998, and I had no intention of continuing to chase John McIntee's pipe dream with him.

McIntee did not give up on the Phoenix Fanjet easily, though. In early 1999, he made a pitch to government officials in Utah to build an assembly plant near the airport in Ogden. U.S. Representative Jim Hansen (R-Utah) told the Salt Lake City newspaper the *Deseret News,* "They have a very compelling case. The thing is they don't have a whole lot of backing at this point." They never got the backing, and the manufacturing plant was never built.

Two years later, Don Jewitt, the Canadian oilman who had been the primary investor in Alberta Aerospace, dissolved the company, only to have it resurrected by McIntee and several other members of the board of directors. Rather than market the plane as an *ab initio* trainer, though, McIntee approached the government of Poland with the idea that the Fanjet could be marketable as a four-seat small business jet. The four-seater had always been in the mind of McIntee, but none was ever produced because of problems getting the two-seater certified.

The last I heard, the Phoenix Fanjet was sitting unused and unloved in a hangar somewhere in Canada.

This was the second time in my career in which a manufacturer tried to get a new aircraft certified but was unable to do so because of a lack of money or a lack of technical expertise within the company. The first was with the Harbin Y-12 in China. The Phoenix Fanjet would not be the last. Another of those projects would come my way some years down the road, after a series of private consulting jobs that had me logging more air miles as a passenger than I ever thought possible.

CHAPTER 14
INTERNATIONAL CONSULTING

Although the Phoenix Fanjet turned out to be a bust because of a lack of funding, my work on that project reinforced my decision to be a private consultant DER rather than a company DER. I enjoyed working as a private consultant because I was working for myself: I was my own boss and could choose for whom I worked, when I worked, and for how long I worked. It was a hell of a lot better than going to the office at 8:00 a.m. and staying until 5:00 p.m., fighting traffic in both directions.

I had formed TopHatter Aviation Consultants, Inc., based out of my home in Georgia after retiring from the FAA. Because of my long experience with the agency and my reputation as a no-nonsense, do-the-job-the-right-way test pilot, I actually had almost more business than I could handle. The FAA and private companies were contacting me with job offers because they knew I was a stickler for ensuring whatever I did would be according to Federal Aviation Regulations.

As mentioned in the previous chapter, my contract with Alberta Aerospace enabled me to take short-term consulting jobs when the Fanjet was down for maintenance. After that contract ended, I spent a lot of time on the road—and in the air. Some of the jobs involved travel to places such as Austria, Italy, Morocco, Saudi Arabia, Singapore, and Spain. Others involved travel to far less glamorous places, such as Springfield, Illinois; and Augusta, Georgia. I made a number of trips to Springfield over the next few years on jobs involving Garrett Aviation, a company that performed a broad range of maintenance services on corporate aircraft.

All those jobs—with one notable exception that I will go into in more detail in a bit—paid well. On most jobs in the continental U.S., I usually spent a day or two at the site and by the third day, I'd be home. But my contracts were written so that if I worked on a Saturday or Sunday, I was paid for it. If I flew to the job site on Saturday or Sunday, even though I wasn't working, the contracts paid while I was in the air flying. My contracts also stipulated that if I flew overseas, I had one day to rest before starting the project.

While I handled the detailed paperwork the FAA required for each project, Becky kept the books to keep track of who paid and when. Most companies paid within a reasonable amount of time after the job was completed. There was one company in Ohio, though, that was notorious about waiting until the 60th day to cut a check to reimburse me for my expenses. I asked them why reimbursement took so long, and they said that after I sent them my expenses, they would send the invoice to their New York headquarters, and New York would send it to a company in Europe that paid their bills. That company would then issue the check and send it back to New York, and New York sent it to the company in Ohio. Finally, the company in Ohio would send it to me. Those delays did not help the cash flow because the credit card companies always wanted their money on time.

The only time I was ever stiffed on a project was in May 1999 when I went to Saudi Arabia to test fly a Boeing 707 after a TCAS II system was installed on it. That plane was part of a fleet of private jets that served Saudi King Fahd bin Abdulaziz Al Saud.

After arriving at the Al Salam Aircraft Company at the international airport in Riyadh, Saudi Arabia, I was taken on a brief walk-through of the plane. This was not your ordinary 707. It was outfitted as a king's plane should be, with chandeliers and gold-plated fixtures throughout. My guide showed me the king's seat and asked me if I wanted to sit in it.

"Hell, no!" I exclaimed. "*I* want to live until tomorrow!"

The testing was relatively simple, and I was in and out of the country in just a few days. After arriving home, I sent an invoice to Al Salam Aircraft Company for consulting fees and expenses that totaled more than $8,500. When I did not hear back from the company in a reasonable amount of time, I started sending letters and Faxes inquiring about my money. The company eventually approved only about $6,100 of my fees and expenses, including a chintzy $5.88 for several days of meals, taxis, parking, visa fees, and other expenses.

The company was partly owned by Boeing and although I sent a strongly worded letter to Boeing expressing my frustration, I never received an answer — or the money I was owed. Needless to say, I never again entertained any job offers from Al Salam Aircraft. As rich as the Saudis were, they were some of the stingiest people for whom I have worked.

Some companies paid by check, others by direct bank deposit. At least one company paid in cash, which caught me a bit by surprise. That was in Spain in early 1998. I went there to perform some flight tests on a de Havilland Canada DHC-4 Caribou that was being converted from a cargo/transport plane to a water bomber to fight forest fires.

The Caribou is a high-wing, twin-engine prop plane with short takeoff and landing capability, often used to ferry cargo or troops into remote areas. This plane was being modified to accommodate an aerial discharge of up to 1,000 gallons of water while airborne. The tests were to determine whether the loading and unloading of the water would adversely affect the performance of the plane and what sections of the Civil Air Regulations might be affected by the modifications.

The tests went relatively smoothly, and the plane eventually was certified for use as a water bomber. As I was getting ready to leave Madrid after finishing my assignment, one of the company officials asked me for a bill. I quickly scribbled out an invoice, which was about $5,000 for my expenses and consulting fee.

The official looked at the invoice, then turned to his assistant and said, "Go get the money for him."

I was expecting the assistant to come back with a check but after a few minutes, he returned with a large wad of $20 and $50 bills in U.S. dollars. I was not sure what to do with it because it was quite a chunk of money, and I did not feel comfortable just stuffing it into my carry-on bag. So, I began folding the bills as small as I could and crammed them into my money belt. I had quite the spare tire flying back to Atlanta.

When I flew overseas for jobs, I always had it written into the contracts that I would fly first class. There's nothing more tiring than flying halfway around the world in a coach-class seat and then being expected to go to work the next day, or even the day after. It is not good for the brain or the body.

The companies for which I consulted usually did not mind that I flew first class, but it tended to rankle FAA officials who occasionally called me about jobs they needed performed because they did not have enough of their own pilots to do them. One day I received a call from someone at the FAA asking if I would be interested in a job in Singapore. The job was relatively simple: flight testing a Boeing 747 for Singapore International Airlines following installation of a passenger entertainment system. This was in 2003 and many airlines that had international flights were upgrading their in-flight entertainment systems.

What a test pilot is looking for in tests of this nature is whether the new system interferes with the radio or any of the cockpit functions, such as the navigation system. You also want to make sure it does not interfere with the plane's internal crew communications system and that you can override the new system if there is a problem.

Most of these tests were not the type you could do in a simulator because a simulator only simulates things; it doesn't necessarily tell you how the airplane will actually react when it's in the air. Depending on the project and how big it was or how new it was, you might do a lot of ground testing; in other cases, you don't have to do much testing before you recommend certification.

On this project, the FAA representative asked me to send him a list of projected expenses for the trip to Singapore, so I put together a list and sent it to him. Of course, it included round-trip first-class airfare.

He called me back and said, "Why in the hell do you pilots all think you have to fly first class?"

"Well," I told him, "we don't have to fly first class. But, if *I* don't fly first class, *I* don't go."

I got a first-class seat.

Not long after that, the International Civil Aviation Organization offered me a full-time job with Singapore International Airlines in Singapore. The job came with a house, a cook, a gardener/watchman, and a full-time bodyguard.

When I told Becky about it, she asked why we needed a bodyguard. I told her I didn't know for certain, although I suspected that as Americans we might be considered targets for terrorists or other unsavory elements. She did not like the idea of having to live some place where we needed a bodyguard, so I turned down that offer. Flying as a test pilot is dangerous enough. There's no need to add a risk on the ground.

One of the things you try to do as a test pilot when you are running tests on a new airplane, or a new system, is ensure that the tests are performed with healthy margins of safety. That is especially true when dealing with systems that assist the pilot in flying under IFR. You don't want to perform these tests under actual IFR conditions because if the system fails, it's a good bet there will be no more airplane and no more pilot.

One test I did on an airplane in Alaska involved a new ground proximity warning system. In order to test the system, we had to fly against a mountain. When I was in the FAA, we would fly those tests against mountains in Arkansas where we had a graphic of the mountain, so we knew the gradient of the slope. As you approached the mountain, the equipment started seeing the gradient, and it told you what the elevation was. Eventually it got to the point where it gave you a warning to break off the flight.

We did not have a graphic for the Alaska test, and on the day we were supposed to perform the test, the mountain was shrouded in fog. Still, the other pilot said, "Let's go anyway."

"Hell, no!" I said sharply. "We're not flying against that mountain in IFR! We can't see the damned thing! We're testing an airplane with equipment that hasn't been approved! No thanks. I want to live longer than that."

To provide some sense of the amount of travel I did during my years as a private consultant, in 2000 I had several projects that required me to fly around the world in a week. The Sunday before Thanksgiving that year, I flew from Atlanta to Paris and from Paris to Morocco, where I did a project. After finishing the project in Morocco, I flew back to Paris to catch a flight to Guangzhou, China, just a bit north of Hong Kong in the southern part of the country. I did a second project there before getting on another plane and flying to Los Angeles, where I caught another flight to Atlanta.

I got back home about 6:00 a.m. the Saturday after I had departed. It was imperative that I get back because our granddaughter Hannah was being christened, and I did not want to miss that event. I made the christening, but by the time the ceremony and the reception were over, I was running on fumes.

In another year as a consultant, I had over 150,000 miles flying just on Delta Airlines, and that was not the only airline on which I flew. Here's just a brief example of some of the jobs I did and some of the places I traveled from 1998 through 2003:

1998

- June 24—Palma Mallorca, Spain. SpanAir DC-9, way-point light failure.
- August 25 — Rome, Italy. Air One Airlines Boeing 737, installation of a single Canadian Marconi CMA-900 Flight Management System with GPS and a single Collins 442 DME, provision for a dual FMS/GPS.
- September 4—Springfield, Illinois. Learjet 35A, TCAS I installation; Gulfstream G-IV, dual Honeywell RCZ-852 Mode S transponders and a Honeywell TCAS 2000 Collision Avoidance System; Learjet 35A, installation of a TCAS I

- December 20—Addis Ababa, Ethiopia. Ethiopian Airlines Boeing 757-200, installation of a Rockwell Collins TCAS System.

1999

- March 2—Coral Gables, Florida. Boeing 737-200, installation of a single Trimble 2101 I/O GPS System.
- March 3—Edwards Air Force Base, California. NASA Dryden Flight Research Center, McDonnell Douglas DC-8-72, installation of a Collins dual-mode S transponder with TCAS.
- May 26—Greensboro, North Carolina. LearJet 25D, installation of BF Goodrich Avionics Systems, Inc., TCAS 791/A, TCAS I, and RGC 250 Radar Graphics Computer.

2000

- March 20—Seoul, South Korea. Asiana Airlines Boeing 747-400, installation of a passenger entertainment system.
- March 31—Taichung, Taiwan. Mandarin Airlines Fokker F27 Mark 50, installation of an Allied Signal CAS 67A TCAS, along with a single Mode S transponder.
- June 28—West Palm Beach, Florida. Canadair CL-600, installation of executive interior and entertainment system.

2001

- February 12—Augusta, Georgia. Dassault Aviation Falcon 900EX, installation of a satellite communications system.

- June 20—Madrid, Spain. Iberia Airlines Boeing 747-400, installation of passenger entertainment system.
- December 6—Cartersville, Georgia. Gulfstream G-159, installation of an Allied Signal CAS-67A TCAS II.

2002

- June 11-27—Daytona Beach, Florida. Mooney M20C, installation of a Power Flow Systems, Inc., extractor exhaust system.
- November 16—Springfield, Illinois. Raytheon Aircraft HS.125.700A, installation of a Honeywell MK-VII Enhanced Ground Proximity Warning System (EGPWS).

2003

- June 9—Singapore. Singapore International Airlines Boeing 747-400, installation of a passenger entertainment system.
- July 2—Pusan, South Korea. Korean Airlines Boeing 747-400, installation of a passenger entertainment system.
- September 25-26 Tel Aviv, Israel. Piper PA-31T, elevator control/trim cable re-routing required to accommodate the camera well in the fuselage belly.

Of course, these were not all the jobs I did during those years; they are just examples of how much travel was involved and how varied the projects were. Interspersed among these dates were numerous other projects throughout the world that kept TopHatter Aviation Consultants, Inc., which was basically Becky and me, constantly working.

In addition to the travel, each job required a significant amount of paperwork to justify my recommendation to the FAA as to whether the modification should be approved or rejected. Sometimes individual reports—known as Type Inspection Reports (TIR) in FAA parlance—consisted of dozens of pages with minute details about how, when, and where the tests were performed and what the outcomes were.

Here is an example of the introductory paragraph to one of my reports:

> This report covers the certification requirements for the installation of an Airshow system within a modified vertical stabilizer on a Boeing B-727. Complete details of the vertical stabilizer modification and the system installation are contained in other reports that will be submitted by the applicant. This report is limited to the evaluation of the aircraft handling qualities and the vibration and buffeting characteristics of the aircraft before and after the installation of the Airshow system. This report does not address airspeed calibration, the flutter margin substantiation, or possible effects on the aircraft icing certification. The external configuration change to the vertical stabilizer appears to be significant, and no data are available from Boeing for the aircraft with the modified vertical.

The report went on to provide details on conditions for the tests and the procedures I employed, such as trimming the aircraft in level flight at or near 40,000 feet at an airspeed close to the maximum certified, accelerating to just under maximum speed, and reducing power and descending. All of it was designed to determine whether the modification to the vertical stabilizer produced significant changes in the aircraft's performance.

Despite the detailed nature of much of the work, it never got old and it never got boring because I was doing everything I had ever wanted to do: fly airplanes for a living. I was approaching my mid-60s in 2003 but had no intention of slowing down. I enjoyed flying too much, and felt I still had a great deal to contribute to the aviation industry.

Still, the constant travel was beginning to wear on me and when an opportunity came along for something that promised to keep me closer to home, I decided to take a chance on it. Besides, it looked like something new and challenging. Little did I know that this would be one of the most frustrating projects of my career, far surpassing the disappointment of the Phoenix Fanjet fiasco.

CHAPTER 15
THE MULTI-MILLION DOLLAR
HONDAJET BOONDOGGLE

In the fall of 2003, I received a call from an old friend, Rich Gritter, who was working for a company in Greensboro, North Carolina, called Atlantic Aero. Rich said the company was about to get a substantial contract with Honda Motor Company to test and certify an airplane the Japanese auto maker had been trying to develop for several years. He asked if I was interested in signing on as a test pilot in the program. He said Honda planned to market the plane as a small business-class jet to compete with similar aircraft being produced by Cessna, Gulfstream, and Embraer.

I was familiar with Honda's reputation for manufacturing and marketing quality cars and motorcycles but had only heard rumors through the aviation grapevine about the company's interest in aviation. Much of what Honda officials had been doing to develop an airplane had been kept under tight wraps in part because of the intense competition in the business-jet market and, in part, because that was just the way the company did business.

As Rich laid out what few details he had about the program, I could not help but think back to my less-than-satisfactory experience with the Phoenix Fanjet a few years earlier. Here was another experimental, one-of-a-kind aircraft being developed for a specific market by a company that had little experience in the aircraft industry. The big difference was that while the Phoenix Fanjet did not have much financial backing, the Honda program had extremely deep pockets.

As a favor to Rich, and because I was curious about what sort of aircraft Honda was trying to develop, I agreed to fly to Greensboro in December 2003 to interview for the test pilot position. The interview was with Michimasa Fujino, the first president of the Honda Aircraft Company and designer of what then was simply called the HondaJet. Fujino was in his early '40s but looked about 10 years younger. He was quick to point out that he had a PhD in aeronautical engineering from the University of Tokyo, but it took only a small amount of questioning by me to get him to admit he was not a pilot and really did not have a great deal of knowledge about what it would take to get an experimental aircraft certified in the United States. He knew what the regulations said, he just did not know how to get there.

During the interview, Fujino revealed few details about the airplane, only that it was going to accommodate four passengers and two pilots, that it would be able to fly as high as 41,000 feet, and that it would be more fuel-efficient flying at altitude than other business jets because of its unique design. But he did not seem to know much more about the airplane—at least he was not revealing any more at that point—because it had only flown once, on December 3, 2003, just before my interview.

Despite some trepidation on my part, when Fujino offered me the job I accepted. It meant I would be working out of Greensboro for an extended period of time and not dashing from city to city around the country and the world as I had been since going into private consulting following my retirement from the FAA. The position with Honda paid well and provided a measure of stability in my life for however long the job lasted. I rented an apartment in Greensboro, and Becky was able to fly up frequently from our home in Fayetteville, Georgia.

I got my first look at the airplane during that visit to Greensboro when I interviewed with Fujino. It was a nice-looking business jet, sort of a shorter, stubbier version of the Cessna Citation CJ4. What set the HondaJet apart from other business-class jets was the engine placement: Instead of being on either side of the fuselage near the tail, the two turbofan engines were placed on struts above the wing. The intent, I later learned, was to reduce engine noise in the passenger compartment and increase fuel efficiency.

Fujino said having the engines above the wing reduced the spanwise flow of air. With most airplanes, as the air comes over the wing it tends to go outboard. Fujino said having the engines above the wing would stop that spanwise flow and keep the air going straight over the wing, thus making the aircraft more efficient due to a reduction in drag. That was his theory. I was never given access to any of the wind tunnel data, so I don't know if that's how it works. Nor do I have a PhD in aeronautical engineering. I just know how to test and certify planes. But, from an aesthetic standpoint, the passengers can't see a danged thing because you've got an engine right outside the window.

As I dug into the history of the development of the HondaJet in later years, I learned that the plane had been in the works since the mid-1980s, when the company assigned Fujino to its fledgling aviation division. Despite his degree in aeronautical engineering, Honda had originally hired Fujino as a research and development engineer in its automotive division. He first worked on anti-lock braking systems for cars before going on to help develop the high-end Honda Acura NSX, which in 2023 had a retail sales price just north of $170,000.

When Fujino was transferred to Honda's aviation division, the company decided to move that operation to the United States to get around the numerous restrictions in Japan on development and testing of experimental aircraft. Besides, the market for small business jets was much greater in the U.S. than in Japan, and it would be much easier to lure customers here than it would be if the company's manufacturing plant and showroom were in Japan.

Fujino's vision was that the HondaJet would do for the business-jet market what the Honda Civic did for the automobile market, but getting there would become a long and incredibly expensive process.

The first plane Honda built was a double-engine turboprop designated the Honda MH02. It was fabricated at Mississippi State University's Raspet Flight Research Laboratory in Starkville, Mississippi, in 1992 and had about 170 hours of flight time. It was never intended as a commercial product; instead, it was used primarily to test the above-the-wing engine mounts and some new composite materials in the fuselage. I never found out how much Honda spent on that project, but estimates were in the tens of millions of dollars. In 1993, the MH02 was shipped off to Japan, never to fly again.

By the time I joined the program in 2003, the HondaJet project was almost two decades old and still had only one experimental aircraft—not counting the MH02—to show for the time invested and money spent. But, as I was to learn, that was the way Honda did business when it was developing a new product: slow and slower with no real concern about how much it cost.

The HondaJet prototype was not a particularly sophisticated

airplane. It was light and nimble and easy to fly but did not have a lot of bells and whistles. It was kind of the Honda Civic of aircraft, only a bit better-looking.

Rich Gritter and I took our first flight in the HondaJet on January 16, 2004. My records from that flight indicate we were testing handling qualities with a rudder modification, evaluating landing gear retraction and extension, and checking out the Garmin 1000 integrated flight instrument system. There were some minor issues with those tests, but nothing serious.

Over the next few months, as we continued to test the airplane, a pattern emerged that made me begin to wonder if Fujino had any idea what it took to get the airplane certified or if he was being restrained by the strict corporate culture of Honda. If Fujino had a plan for the plane, he would not let Rich and me in on it.

When most aircraft manufacturers come out with a new model, they generally have a few available for testing. They just pull out the paperwork and say, "Here's what we do in the first test, here's what we do in the second test, here's what we do in the third test," and on down the line. As bugs are found, the engineers and designers iron them out in the model just tested, and then fix the problems in the yet-to-be-tested models. Honda had only one HondaJet and no clear, detailed plan for testing it—akin to trying to play darts blindfolded. Fujino could not show me a plan—if he had one—that would have led us in the direction of certification.

Instead, the tests were scattered all over the place and many of them served no useful purpose. For example, one of the first tests you should perform on any new model aircraft is the one that determines the stall speed. You must determine at what speed that new airplane is going to stall, particularly on jets, before you are able to determine the other V speeds that go into the certification process. Fujino never wanted us to do a stall test; in the three years I flew that airplane, Fujino would not authorize a stall test. As a result, we were never able to get any of the other V speeds.

Once when I did a power-off landing, I thought Fujino was going to have heart failure on the spot. The engines were actually at idle, but we were having some engine problems and I told Rich we needed to find out what the heck would happen if we lost engines on landing. Fujino chewed our butts out when we got back, but at the time we just assumed he was being very, very conservative about the airplane.

Fujino would not even let us do an engine cut on takeoff.

"What do we do if it happens for real?" I asked him.

"We'll cross that bridge when we get to it," he said.

"That's a poor time to cross that bridge!" I snapped back.

Despite what Fujino did not want, at my insistence Rich and I did a number of engine cuts on takeoffs and a number of simulated power-off landings to a full stop.

I tried to explain to Fujino that the HondaJet was a toy compared to an F-104. I told him I had done all kinds of simulated power, no-power landings in an F-104. He seemed surprised. I told him it was at Edwards Air Force Base, which is at 2,000 feet elevation. I said we started at 25,000 feet and lost 23,000 feet while making a 360-degree turn.

He looked as if he was in shock and asked incredulously, "They let you do that?"

"It's required," I answered. "It's no big deal," I added, trying to reassure him that my thousands of hours as a test pilot would guarantee no harm would come to the HondaJet.

When I would ask him about doing some of the tests to determine various V speeds and how important they were for the certification process, he would just hem and haw and say, "We're not ready for this" or "We're not ready for that."

"Fujino," I said with as much patience as I could muster, "are you aware that stall speeds are some of the first bits of data a company gets on a new jet airplane?"

"Yes, but we're not ready for that," he said dismissively.

Fujino had no idea about how to get the V speeds and did not want to know what they were. The easiest way to avoid finding out what those numbers are is to not put them on the test card. If they are not on the card, you don't do the tests, and they were never on our test cards.

Most of the tests we did were for range and speed performance, maximum speed at different altitudes, and range at different altitudes. Eventually Fujino was going to need those, but he did not need them at that stage of the program. Those are about the last things you need, except to convince potential buyers you are going to have the range you are advertising. We also did some tests for handling qualities, but they were very minimal. It was not an in-depth test program by any stretch of the imagination.

Fujino was not making any major changes to the plane after our tests. He seemed to have little interest in what we learned. Instead, he was constantly focused on how much the plane weighed before and after every flight. It struck me as being rather odd, but Fujino insisted that after every flight we put the plane on the scales to get what is known as "weight and balance." I told him we knew the weight and balance before we took off and if they filled it with the same amount of fuel after the flight, the weight and balance would be the same. They did not have to put it on the scales to measure weight and balance. Yet, after every flight they would tow the plane into the hangar, and we would waste two hours while they re-weighed the plane. The only thing that changed after every flight was the jet fuel, but I could never make Fujino understand that concept. Why he worried so much about weighing the HondaJet after every flight, I'll never know. Maybe it was a Honda thing.

My sense after a few months of this was that Fujino did not want the more serious and specific tests conducted because he was afraid of losing the airplane. Nobody wants to lose an airplane, but that's part of flight testing. You have to put in some safety requirements to reduce the possibility that you'll lose an airplane, but you can't rule out that you'll never have an accident. That was not Fujino's way of viewing the situation, however.

The fact that there was only one aircraft was not my problem. Rumors were that Honda already had many millions of dollars tied up in the program, so spending another million or so to build a second plane seemed like peanuts to me, especially for Honda.

I came to believe Fujino either did not have confidence in what he was doing or was being restrained by Honda's restrictive top-down corporate culture. Fujino thought what Rich and I were doing to get the aircraft certified was too dangerous, so he did not want us to run the tests that test pilots normally perform on new aircraft. He was not a risk-taker, as you sometimes have to be with a new aircraft. He had absolutely no ability or desire to think outside the box. If it was not written down by him, you by-God better not do anything that deviated from his list.

Fujino seemed primarily concerned about being able to tell a customer he could get 700 miles on a tank of gas, could fly to 40,000 feet, could carry six people including two pilots, and that the airplane looked good and had low drag because the engines were above the wing.

Since Honda had already invested a sizeable amount of money in the project, once Rich and I started testing, the company seemed to go out of its way to pull back the curtain that had been shielding the plane's development for years. Fujino brought in reporters from various aviation magazines to extoll the virtues of the HondaJet and compare it to other small business jets, even though we were a long, long way from certification.

In most of those reports, whatever flaws we were discovering in the HondaJet were glossed over while Fujino was portrayed as something of a visionary because of the plane's unique design and his use of composite materials in its fabrication. Those aviation journalists never understood they were being taken for a ride, figuratively, in an effort to market the plane even before it went into production.

Honda even released a promotional video for the plane that included a snippet showing Rich and me riding Honda motorcycles out to the plane for one of our tests. That was about 2004 or 2005 but even then, the HondaJet was still more of Fujino's dream than a marketable airplane.

In July 2005, Honda decided to take the plane public, so Rich and I flew it to Oshkosh, Wisconsin, for display at the Experimental Aircraft Association AirVenture show. Among the reporters present were two from *Bloomberg News*. They wrote a story about the HondaJet, and the article, along with a photo of me standing in front of the aircraft, ran on the front page of the *Seattle Times* business section on July 30, 2005.

Fujino was at the air show as well and was the perfect corporate spokesman for Honda. "Aviation has long been a dream for Honda," he told reporters, "and the HondaJet is the embodiment of that dream."

That dream was still little more than a dream, though. By the time of the Wisconsin air show, I was beginning to think it was time to get out of Dodge. I was not at all confident that Fujino would be able to get the HondaJet through certification in a reasonable amount of time. Even though he had the money at his disposal and the time to get the necessary tests done, it was just that he did not want to do the tests.

I hung on with Honda for another year, but there was little progress with the testing. I knew the end had finally come when Fujino brought in two pilots from the civilian test pilot school in Mojave, California, to do the stall speed tests. He told Rich and me that we were being too loose with the airplane and were trying too many things he thought were dangerous. I pointed out to him that Rich and I had flown the plane for a number of hours but had never damaged the aircraft. He did not care. He apparently wanted someone to fly the airplane over whom he had more control.

Instead of doing the types of tests to determine stall speed with which Rich and I and virtually every other test pilot was familiar, Fujino set up a program that did the stall speed tests incrementally. The Honda crew would limit the angle of attack to something below what they thought was the stall angle of attack. They approached stall speed a quarter of a degree for each test flight. They were trying to sneak up on the stall speed on tiptoes. I knew at that rate it would take a year or more to get all the flight data required for certification.

Technically, I was still on the payroll when these two new pilots came to work, but I left shortly after that because it was pretty obvious I would not be around much longer, even though Fujino never said anything to me about leaving. I was not angry, but I was incredibly frustrated by the process. I think if Fujino had allowed Rich and me to perform the flight testing we normally did for certification, we could have had that airplane certified by 2006 or 2007. But Honda and Fujino chose not to allow us to perform the certification tests as we had been trained to do.

I lost track of the HondaJet certification process once I left the company in the summer of 2006. Apparently, though, it continued to move along at something slower than a snail's pace. Whatever they were doing, it was not until December 2015, nearly three decades after Honda had embarked on getting a small jet on the market, that the HondaJet received its type certification from the FAA.

Exactly how much Honda spent building the HondaJet and getting it certified is known only to company officials. Reuters news service quoted one aviation industry source as saying it was roughly $1 billion. *Forbes* magazine estimated it cost Honda between $1.5 billion and $2.5 billion to get the small jet certified. I had heard that it was in the neighborhood of $1.6 billion.

Whatever the cost, Honda and Fujino had their jet, officially designated the HA-420 HondaJet. The sales price is about $5 million per plane, out of which Honda hopes to realize a profit of $1 million.

At that rate, it will take more than 1,000 HondaJet sales to get some glimpse of a profit from the project.

EPILOGUE

Reflections on a Pilot's Life

After leaving the HondaJet project in the summer of 2006, I went back to being a consultant and test pilot for various private aviation firms that needed certification on aircraft and systems the FAA did not have the resources to handle. The FAA bureaucracy had become too big and bloated and unwieldy—and seemed to be getting much more so by the month. There were more paper pushers than test pilots. The agency had too many tests to do on too many aircraft and too few pilots to do them, but it had plenty of bureaucrats who didn't know an aileron from an altimeter.

Airplanes have become very complex because so many of their functions are now handled by computers, so I can understand the argument that says the U.S. government cannot fund enough people, especially enough engineers, to do the all the testing on an airplane, especially a new airplane like Boeing's 737 MAX.

But ensuring the public's flying safety is the FAA's reason for existing. It is supposed to oversee the airlines and aircraft manufacturers. It should be incumbent upon the FAA to have its test pilots and engineers thoroughly test all new systems, new aircraft, and new procedures. It should be mandatory for FAA pilots to conduct the tests, even after the company has done its own testing. The FAA has the prerogative to do as much or as little testing as it desires and should not be influenced by what the company DERs tell them.

What has been the norm for years, however, is that when it comes to the Boeings and the Cessnas and the Beeches, the attitude is: "Leave them alone; they know what they are doing." In the case of the 737 MAX, some folks *didn't* know what they were doing. Or they *didn't* listen to someone who knew what they were doing. What happened with that airplane and the problems it had did not surprise me one little bit because of the system that produced it.

Will there ever be changes to the system that will better ensure the safety of the flying public?

I don't think so, because that is how "the system" works, and changing something so ingrained is almost impossible, especially when so many politicians and so much money are involved. I did what I could to fight against that system when I worked for the FAA, which did not earn me a lot of friends among the upper echelons of the FAA bureaucracy, but I never really cared about that. I was more concerned about doing the right thing at the right time and doing it well.

I held myself to a certain standard when it came to testing airplanes, and I had little patience with anyone who did not set the same high standards because by ignoring those high standards, they put people's lives at risk.

Do things the right way or don't do them at all. That's the principle by which I lived and worked and the principle by which I expected others to do their jobs. I learned that from my father and had it reinforced during my time at the Naval Academy. I have also tried to stress this principle in this book: that whatever I did when it came to testing airplanes, I did it the way it should be done, not the way that was easiest for me.

As rigid as I was about doing tests the right way, I was not so focused on not making a mistake that I did not enjoy flying airplanes. I had set a goal for myself early in life and realized that goal—I was paid to fly airplanes. Whether it was a Beech 300 or a Boeing 767, the enjoyment of flying was something that never got old, no matter how many times I did it.

I've lost track of all the airplanes I've flown over the years, but as I have recounted in previous pages, I flew a number of military aircraft, starting with the single-engine T-34B Mentor in flight school and moving on to the F-4 Phantom in Vietnam and from there to the experimental YA-9 jet while I was with Northrop.

In later years, I was certified as an air transport pilot in the following civilian aircraft:

- Boeing 727
- Boeing 757
- Boeing 767
- Beech 300
- Cessna 500
- Convair 340
- Convair 440
- DC-9
- Embraer 120
- Gulfstream 450
- Gulfstream IV
- North American Rockwell NA-265 Sabreliner

I also had commercial flight privileges in two types of gliders, a rating in hot air balloons, and had a flight engineer rating in turbojets.

In 2009, shortly before I turned 71 years old and after about 47 years of flying, I decided it was time to come down from the clouds and retire. I gave it up grudgingly, and still a day does not go by that I do not miss the excitement of getting into the cockpit of an aircraft, going through the pre-flight checks, revving up the engines, and getting that plane off the ground and into the air. Flying was my livelihood, but it was also my dream and my passion.

In writing this book, I hope some of that passion and enjoyment comes through. I also hope that some of my experiences with the FAA over the years can help shed some light or put some perspective on what the agency has become today.

If there is one thing I want people to take from this record of my life it is that *I lived my dream and I lived it on my own terms.*

General Douglas MacArthur once said, "Old soldiers never die; they just fade away." It's a bit different for those of us who were aviators: "Old pilots never die; they just fly away."

ACKNOWLEDGMENTS

The publication of this memoir would not have been possible without the perseverance and dedication of Ron Martz, whose thorough research, careful editing and inherent creativity transformed a series of interviews into a informative account of a memorable life. Appreciation is expressed to Tom Peters, Dave Gollings and Cindy Gollings, who shared recollections of the Atlanta Aircraft Certification Office when Dave served as the chief of the Flight Test department. Appreciation is also expressed to the employees of Deeds Publishing, Inc., of Athens, GA, whose professional standards have led to the publication of this memoir. Thanks is also expressed to son David Paul West, II, whose expertise in the field of aviation provided clarity in areas of avionics.

ABOUT THE AUTHOR

David Paul West was a naval aviator and served aboard the *USS Roosevelt* during the Vietnam War. Later, he became a test pilot. His career in aviation spanned almost 47 years. He began his memoir in 2019, but the project was put on hold during the COVID pandemic. He died in July 2021 from complications following a fall at home. His widow, Rebecca West, began the process of completing the memoir with the assistance of Ron Martz, a noted author on military topics.

ABOUT RON MARTZ

Ron Martz is the co-author of five books on military history and national security issues, including *Disposable Patriot: Revelations of a Soldier in America's Secret Wars* (with Jack Terrell, National Press Books, 1992); *Solitary Survivor: The First American POW in Southeast Asia)*, (with Col. Lawrence R. Bailey, Jr., Brassey's, 1995); *White Tigers: My Secret War in North Korea*, (with Col. Ben S. Malcom, Brassey's, 1996); *Heavy Metal: A Tank Company's Battle to Baghdad*, (with Capt. Jason Conroy, Potomac Books, 2005); and *Survival: How a Culture of Preparedness Can Save You and Your Family from Disasters*, (with Lt. Gen. Russel Honoré, Simon & Schuster, 2009). During his 40-year newspaper career he reported from more than 35 countries while covering seven wars and regional conflicts. He was nominated for the Pulitzer Prize in 2003 for his reporting on the war in Iraq. A veteran of the U.S. Marine Corps, he lives in Cumming, Ga., with his wife, Mary, a technical writer, and their four rescue dogs.

www.ingramcontent.com/pod-product-compliance
Lightning Source LLC
Chambersburg PA
CBHW021708120626
46545CB00004B/1465